QH

SO-ATF-646

2007

CCU Library
8787 W. Alameda Ave.
Lakewood, CO 80226

The International Society for Science & Religion was established in 2002 to promote education through the support of inter-disciplinary learning and research in the fields of science and religion. Our current membership of 140 comes from all major faith traditions and includes non-religious scholars. Inducted by nomination only, they are drawn from leading research institutions and academies worldwide. The Society embraces all perspectives that are supported by excellent scholarship.

In 2007, the Society began the process of creating a unique resource, *The ISSR Library*, a comprehensive reference and teaching tool for scholars, students, and interested lay readers. This collection spans the essential ideas and arguments that frame studies in science, religion, and the human spirit.

The Library has been selected through a rigorous process of peer review. Each constituent volume makes a substantial contribution to the field or stands as an important referent. These books exhibit the highest quality of scholarship and present distinct, influential viewpoints, some of which are controversial. While the many perspectives in these volumes are not individually endorsed by the ISSR, each reflects a facet of the field that is worthy of attention.

Accompanying the Library is *The ISSR Companion to Science and Religion*, a volume containing brief introductory essays on each of the Library's constituents. Users are encouraged to refer to the *Companion* or our website for an overview of the Library.

Pranab K. Das II, Executive Editor

Editorial Board
John Hedley Brooke
Philip Clayton
V. V. Raman
Wesley J. Wildman

Christopher Knight, ISSR Executive Secretary
Ronald Cole-Turner, ISSR Board Liaison

For a complete list of Contributing Editors, please visit
www.issrlibrary.org/contributingeditors

For a complete list of Library Selections please visit
www.issrlibrary.org/selections

DARWIN'S GIFT
TO SCIENCE AND RELIGION

DARWIN'S GIFT
TO SCIENCE AND RELIGION

FRANCISCO J. AYALA

Joseph Henry Press
Washington, D.C.

Joseph Henry Press   **500 Fifth Street, NW**   **Washington, DC 20001**

The Joseph Henry Press, an imprint of the National Academies Press, was created with the goal of making books on science, technology, and health more widely available to professionals and the public. Joseph Henry was one of the founders of the National Academy of Sciences and a leader in early American science.

Any opinions, findings, conclusions, or recommendations expressed in this volume are those of the author and do not necessarily reflect the views of the National Academy of Sciences or its affiliated institutions.

Library of Congress Cataloging-in-Publication Data

Ayala, Francisco José, 1934–
 Darwin's gift to science and religion / by Francisco J. Ayala.
     p. cm.
 Includes bibliographical references and index.
 ISBN-13: 978-0-309-10231-5 (cloth cover)
 ISBN-10: 0-309-10231-6 (cloth cover)
 ISBN-13: 978-0-309-66174-4 (pdf)
 ISBN-10: 0-309-66174-9 (pdf)
 1. Natural selection. 2. Evolution (Biology) 3. Intelligent design. 4. Darwin, Charles, 1809–1882. I. Title.
 QH375.A93 2007
 576.8—dc22

                              2007005821

Cover design by Michele de la Menardiere
Cover illustration "THINKER" POSE © James Balog/Getty Images

Copyright 2007 by Francisco J. Ayala. All rights reserved.

Printed in the United States of America

To Hana,
with love

Contents

PREFACE

The message that this book conveys can be simply stated:
Science and religious beliefs need not be in contradiction. If
they are properly understood, they *cannot* be in contradiction
because science and religion concern different matters. Science
concerns the processes that account for the natural world: how
the planets move, the composition of matter and the atmo-
sphere, the origin and function of organisms. Religion concerns
the meaning and purpose of the world and of human life, the
proper relation of people to their Creator and to each other,
the moral values that inspire and govern people's lives.

The proper relationship between science and religion can
be, for people of faith, mutually motivating and inspiring.
Science may inspire religious beliefs and religious behavior, as
we respond with awe to the immensity of the universe, the wondrous diversity of organisms, and the marvels of the human
brain and the human mind. Religion promotes reverence
for the creation, for humankind as well as the environment.
Religion may be a motivating force and source of inspiration

for scientific research and may move scientists to investigate the marvelous world of the creation and to solve the puzzles with which it confronts us.

Along the way, I will belabor two points primarily addressed to people of faith. The first point is that science is here to stay. No matter what flaws or unknowns religious believers may see in scientific knowledge, science will continue its relentless advance toward solving the puzzles of the world of nature. The condemnation of Galileo by the Catholic Church in the seventeenth century did not bring astronomy to a halt. Rather, we marvel at the immensity of our galaxy and the myriad galaxies beyond. Those who see fault with the theory of evolution may seek to keep it out of the school curriculum or to belittle its accomplishments, but the thousands of scientists who in hundreds of universities and other institutions pursue evolutionary research will persist in their endeavors and continue the advance of knowledge. Universities, foundations, and governments will continue investing millions of dollars in evolutionary research, and thousands of research papers will continue being published yearly in scientific journals.

A second point that I make in this book may come as a surprise to people of faith and scientists alike. I assert that scientific knowledge, the theory of evolution in particular, is consistent with a religious belief in God, whereas Creationism and Intelligent Design are not. This point depends on a particular view of God—shared by many people of faith—as omniscient, omnipotent, and benevolent. This point also depends on our knowledge of the natural world and, particularly, of the living world. The natural world abounds in catastrophes, disasters, imperfections, dysfunctions, suffering, and cruelty. Tsunamis bring destruction and death; volcanic eruptions erased Pompeii and Herculaneum, killing all their citizens; floods and droughts bring ruin to farmers. The human jaw is poorly designed, lions devour their prey, malaria parasites kill millions of humans every year and make 500 million sick. I do not attribute all this misery, cruelty, and destruction to the specific design of the Creator. About 20 percent of all human pregnancies end in spontaneous abortion during the first two months. That is 20 million natural abortions every year. I shudder in terror at the thought that some people of faith would implicitly attribute this calamity to the Creator's faulty design. I rather see it as a consequence of the clumsy ways of the evolutionary process. The God of revelation and faith is a God of love and mercy, and of wisdom.

Darwin's theory of evolution is a gift to science, and to religion as well. This book gives my reasons. I hope you, the reader, will find them cogent and perhaps be persuaded.

Irvine, California
March 1, 2007

1

INTRODUCTION

Is God willing to prevent evil, but not able? Then he is impotent. Is he able, but not willing? Then he is malevolent. Is he both able and willing? Whence then evil?

David Hume
Dialogues Concerning Natural Religion, p. 244

The Dominican monastery of San Esteban in Salamanca, Spain, is a complex of buildings dominated by the huge church of San Esteban, a majestic example of late Gothic architecture, designed in the fifteenth century by one of its monks, Juan de Alava. The magnificent Spanish Plateresque façade opens into a severe and uncluttered interior, dominated by the sumptuously ornate and gilded high altar, a 1692 Baroque masterpiece designed by José Churriguera. The main entrance to the monastery is through an Italianate Renaissance portico. Beyond the main Gothic cloister and grand staircase is the *Claustro de Colón*, where in visits during 1491–1492, Columbus discussed with Padre Deza, the scholarly confessor of Queen Isabella, his calculations of Earth's diameter, seeking Deza's endorsement to obtain the financial and political support of the Queen for the expedition that a few months later would discover the Americas and initiate a new episode in human history.

This was the setting in which I first encountered the argument from design in the late 1950s. As a student of theology in the Pontifical Faculty of San Esteban, I learned of the five arguments, or ways, for demonstrating God's existence that had been formulated by Thomas Aquinas (1224–1274) in the *Summa Theologiae*. The "fifth way" is an argument from design, based on the designed purposefulness of the universe: "We see that things that lack intelligence act for an end, which is not fortuitous but results from design . . . directed by some being endowed with knowledge and intelligence. . . . Therefore some intelligent being exists by whom all natural things are directed to their end; and this being we call God." Aquinas's argument was founded on the universe and its parts, all harmoniously fitting together and thus evincing their design.[1]

The argument from design for the existence of God based on the complex organization of living things would be formulated much later by William Paley (1743–1805) in his *Natural Theology* (1802). Famously, Paley compared a telescope and the human eye, arguing that both were designed, one by a telescope maker, the other by the same Power who had also created the immense diversity of organisms, with their elaborate contrivances—eyes, wings, gills—all exquisitely designed for effectively accomplishing the functions of life.

Several years would pass after my theology studies in Salamanca before I learned about Paley's work. This was in 1961, at Columbia University, in New York City, where I began doctoral studies in genetics and evolution. It was an unexpected turn of events for me, coming from conservative Spain, to discover that there was in the United States a strong creationist current that saw Darwin and the theory of evolution as contrary to religious beliefs. In Salamanca, in my theological studies, evolution had been perceived as a friend, not an enemy, of the Christian faith. Evolution, I learned in my theology classes, had provided theologians with the "missing link" in the explanation of evil in the world or, in theological parlance, evolution had solved the "theodicy" problem. A dictionary definition of theodicy is "defense of God's goodness and omnipotence in view of the existence of evil."

Setting the Stage

The problem of evil has been succinctly stated in the Christian tradition as a dilemma: "If God cannot prevent evil, God is not omnipotent; if God can prevent evil but does not, God is not benevolent. But evil exists—how come?" If the reasoning is valid, it would follow that God is not all-powerful or all-good. Christian theology accepts that evil exists, but denies the validity of the argument.[2]

Traditional theology distinguishes three kinds of evil: (1) moral evil or sin, the evil originated by human beings; (2) pain and suffering as experienced by human beings; (3) physical evil, such as floods, tornados, earthquakes, and the imperfections of all creatures.

Theology has a ready answer for the first two kinds of evil. Sin is a consequence of free will; the flip side of sin is virtue, also a consequence of free will. Christian theologians have expounded that if humans are to enter into a genuinely personal relationship with their maker, they must first experience some degree of freedom and autonomy. A virtuous life *earns* the eternal reward of heaven. Christian theology also provides a good accounting of human pain and suffering. To the extent that pain and suffering are caused by war, injustice, and other forms of human wrongdoing, they are also a consequence of free will; people choose to inflict harm on one another. On the flip side are good deeds by which people choose to alleviate human suffering.

What about earthquakes, storms, floods, droughts, and other physical catastrophes? Enter modern science into the theologian's reasoning. Physical events are built into the structure of the world itself. Since the seventeenth century, humans have known that the processes by which galaxies and stars come into existence, the planets are formed, the continents move, the weather and the change of seasons, and floods and earthquakes occur are natural processes, not events specifically designed by God for punishing or rewarding humans. The extreme violence of supernova explosions and the chaotic frenzy at galactic centers are outcomes of the laws of physics, not the design of a fearsome deity. Alas, theodicy still encountered a seemingly insurmountable difficulty. If God is the designer of life, whence the lion's cruelty, the snake's poison, and the parasites that secure their existence only by destroying their hosts?

I first heard about evolution in the Catholic grammar school and high school I attended in Madrid. My first science class, in sixth grade, was taught by Father Pedro, a gentle soul, who would catch fire when explaining science. Evolution was also taught at length in the biology course I attended at the University of Madrid. That the theory of evolution might conflict with the teachings of the Catholic Church, which pervaded life in Spain during the 1940s and 1950s, was never mentioned, as far as I can remember, except perhaps as a whimsical conceit or as a timid concern of intellectual conservatives. Surely, the 1950 encyclical *Humani generis* by Pope Pius XII had put the matter to rest. Biological evolution, the Pope wrote, is compatible with the Christian faith.

Later, when I was studying theology in Salamanca, Darwin was a much-welcomed friend. The theory of evolution provided the solution to the remaining component of the problem of evil. As floods and drought were a necessary consequence of the fabric of the physical world, predators and parasites, dysfunctions and diseases were a consequence of the evolution of life. They were *not* a result of deficient or malevolent design: the features of organisms were not *designed* by the Creator.

Evolution by natural selection is Darwin's answer to Paley. It is also the solution to the last prong of the problem of evil. Theology professors in Salamanca saw in the theory of evolution a significant, even definitive, contribution to theodicy. I was, therefore, much surprised when I became aware of the creationist movement in the United States and the pervasive reservations against the theory of evolution.

Since the 1970s I have been much involved in seeking a solution to the perceived conflict between evolution and religion. I was an expert witness in the Little Rock, Arkansas, trial decided by Federal District Judge William R. Overton in 1982 (*McLean v. Arkansas*) and I participated in the preparation of an Amicus Brief submitted to the Supreme Court of the United States by the National Academy of Sciences and the American Association for the Advancement of Science (*Edwards v. Aguilar*, 1987). These two decisions declare that teaching "creation science" in public schools is contrary to the Constitution of the United States. Over the past thirty years, I have published editorials, critiques, and articles and have delivered innumerable public lectures at civic and religious gatherings of various Christian denominations and at colleges and universities. The message has always been twofold: (1) evolution is good science and (2) there need not be contradiction between evolution and religious beliefs.

I have sought over the years to persuade my readers and listeners that evolution is here to stay, as a well-corroborated scientific theory, but that Christians need not see evolution as a threat to their beliefs. My conviction is that the theory of evolution is theology's disguised friend, not its enemy. This also is the message of this book: There need not be conflict between religion and science. Apparent contradictions only emerge when either the science or the religious beliefs, or very often both, are misinterpreted. So, please read on; hear me out. I'll seek to persuade you. If I fail, you'll have the satisfaction of having listened to my reasons and, perhaps, you may reinforce or clarify the grounds for your objections.

Making the Case

In his *Natural Theology* of 1802, William Paley made the strongest possible case for intelligent design, based on extensive and accurate biological knowledge, as extensive and accurate as was available at the time. Paley made well the case that the human eye is as complex a contrivance as a watch or a telescope, with several parts all required to fit precisely for achieving vision. I'll introduce Paley's persuasive arguments in Chapter 2. He explored the diversity of organs and limbs in all sorts of organisms, precisely designed to accomplish their function. Paley saw that the relationship between mates of the same species, between animals of different species, and between organisms and their environments appeared to have been precisely designed by an omnipotent Creator. In the first half of the nineteenth century, other scientists (such as Sir Charles Bell in *The Hand, Its Mechanisms and Vital Endowments as Evincing Design*) would explore scientific evidence of intelligent design, but the argument from intelligent design has never been made, either before or afterward, as extensively or as forcefully as it was made by Paley.

Paley's and Bell's evidence for design was persuasive and, indeed, decisive based on the scientific knowledge available in the first half of the nineteenth century, but their arguments crumbled after Charles Darwin's discovery of natural selection and the publication of his *The Origin of Species* (hereafter abbreviated as simply *The Origin*) in 1859. Darwin is deservedly credited for accumulating convincing evidence from paleontology and biology that demonstrated life's evolution. In Chapter 3, I make the point that, important as that demonstration was, it was not Darwin's primary concern. Darwin was, first and foremost, motivated to show that his discovery of natural selection provided a scientific explanation of the design of organisms. Darwin's account of biological design implied, as a necessary consequence, that organisms would have evolved through time and diversified in different habitats. Therefore, Darwin collected evidence of biological evolution in order to corroborate his explanation of design by natural selection. I point out in Chapter 3 that Darwin's discovery of natural selection is one of the most significant events in intellectual history, because it completed the Copernican Revolution. The scientific advances of the sixteenth and seventeenth centuries had brought the phenomena of inanimate matter—the motions of the planets in the heavens and of physical objects on Earth—to the domain of science: explanation by natural laws. Natural selection similarly provided a scientific account for the design and diversity of organisms, which had been left out by the Copernican Revolution. With Darwin, all natural processes, inanimate and living, became subject to scientific investigation.

Accounts of natural selection, scholarly or popular, often short-change Alfred Russel Wallace, its independent co-discoverer. I suggest that this neglect is not misplaced. Wallace saw natural selection as promoting evolution, which he saw as progressive, not as the explanation of design, which is its ultimate significance.

Chapter 4 is an attempt to explain natural selection briefly to nonbiologists. I provide a simple definition, well aware that a proper explanation of the process calls for an extensive treatise. That working definition allows me to highlight important features of the process of natural selection: it is grounded on genetic change; depends on spontaneous mutations; is opportunistic, that is, modulated by the past history of organisms and the demands of the environment; and is "creative," so that it gives rise to genuine novelty, organisms and features that are designed for specific ways of life but would never have come to be without natural selection. I provide a simple example, using bacteria, to show how events each of extremely improbable occurrence—suitable mutations—combine and become actuated in organisms. The fauna and flora of the Hawaiian Islands illustrate some dominant features of natural selection: opportunism, adaptation, and prevalence of some kinds of organisms and absence of others well suited for Hawaiian habitats.

I review evolutionary history in Chapters 5, 6, and 7, where I introduce the evidence for evolution with a statement (accompanied by a supporting illustration) that is sure to surprise most readers: gaps in the reconstruction of evolutionary history from all living organisms back to their common ancestor no longer exist. The evidence comes from the recent revolution in molecular biology. Chapter 5 is mostly dedicated to the kind of evidence that was available to Darwin, although made current: the fossil record of organisms that lived long before the present, such as a primitive horse that lived 50 million years ago; comparative anatomy, showing that the forelimbs of humans, dogs, whales, and birds are modifications of ancestral (reptile) forelimbs; comparative embryology and vestigial organs, such as the human vermiform appendix and our minitail; and biogeography, the peculiar distribution of plants and animals that tells us so much about the history, not only of organisms but also of continents and islands. In that chapter, you will find unfamiliar names such as *Archaeopteryx* and *Tiktaalik. Archaeopteryx* fossils have feathers and a skeleton intermediate between birds and their dinosaur ancestors. *Tiktaalik* is the whimsical name for several specimens, described in early 2006, of an animal intermediate between fish and tetrapods (amphibians).

Darwin extended the theory of evolution by natural selection to humans in *The Descent of Man*, which he published in 1871, twelve years after *The Origin*. Intermediate fossils between humans and apes were yet to be discovered—the "missing links" alleged by Darwin's critics. As I explain in Chapter 6, the missing links are no longer missing. Thousands of intermediate fossil remains (known as "hominids") have been discovered since Darwin's time, and the rate of discovery is accelerating. The earliest hominids are about 6 million years old: *Sahelanthropus* from Chad, *Orrorin* from Kenya, and *Ardipithecus* from Ethiopia. Several fossil specimens of *Australopithecus afarensis* have been discovered in the Afar region of Ethiopia; these are likely ancestors of ours that lived about 4 million years ago, were bipedal, but had small brains, about one pound in weight, one-third the brain size of modern humans. *Homo habilis*, our ancestors of 2 million years ago, had one-and-a-half-pound brains. Their descendants, *Homo erectus*, who spread from Africa to Asia and Europe, had two-pound-plus brains; they and their relatives lived for several hundred thousand years. Our species, *Homo sapiens*, evolved in Africa about 150,000 years ago and then spread through the world's continents.

Two major puzzles of human evolution remain. One puzzle is the genetic basis of the ape-to-human transformation. The human genome and the chimpanzee genome have been deciphered. Each consists of about 3 billion letters—the linearly arranged nucleotides of four kinds that make up the DNA. The human and chimp genomes differ by little more than 1 percent and yet we are so different in important ways: much larger brain, language, technology, art, ethics, and religion. The other puzzle is the brain-to-mind transformation. We know that the 30 billion neurons in our brains communicate between themselves and with other nerve cells by chemical and electric signals. How do these signals become transformed into perceptions, feelings, ideas, critical arguments, aesthetic emotions, and ethical and religious values? And how, out of this diversity of experiences, does a unitary reality emerge, the mind or self? The soul created by God, you might say, accounts for both transformations: ape to human and brain to mind. This religious answer may be satisfactory for believers, but it is not *scientifically* satisfactory. I still want to know how the anatomical and behavioral traits that differentiate us from apes emerge out of our genetic differences; I also want to know the biological correlates that account for mental experiences.

Molecular biology emerged as a discipline 100 years after Darwin, following the 1953 discovery of the double-helix structure of DNA, the hereditary chemical. Molecular biology provides the strongest evidence of biological evolution and makes it possible to reconstruct evolutionary history with as much detail and precision as anyone might want (Chapter 7). The chemical components of life and their proportions, DNA, the genetic code that conveys the genetic information from the nucleus to the cell, the twenty amino acid components of proteins and enzymes—they are all the same in all organisms from bacteria and protozoa to plants and animals. This uniformity makes sense only if it is due to a common origin. For reconstructing evolutionary history it is all-important that genetic information is stored in the linear array of letters (nucleotides) that make up the DNA. The DNA sequences from different organisms can be aligned. The number of letters that are different between organisms reflect the time elapsed since their last common ancestor. One reason why molecular evolutionary biology is so powerful is that it allows us to compare the most diverse kinds of organisms, something not possible for comparative anatomy or the fossil record. DNA sequences of humans, flies, trees, and bacteria can be aligned with one another and with all sorts of other organisms in order to ascertain their evolutionary history.

Another reason why molecular biology is powerful is multiplicity. There are thousands of genes in each organism. If the results of one study are not as precise or as detailed as the investigator desires, he can study more and more genes until he reaches the desired accuracy and detail. There is virtually no limit.

Modern versions of the argument from design, "intelligent design" (ID) as it has been currently named, are considered in Chapter 8. As you read this chapter, you will discover that, although I tried, I couldn't find many saving graces in ID. The one saving grace is the proponents' motivation: proponents of ID want to discover God and faith in science. I, on the contrary, hold that religious beliefs should seek justification on the solid rocks of faith and revelation, not on scientific knowledge—which by its very nature is never definitive or forever valid. ID proponents say that evolution is "only" a theory. But "theory" is a term used by scientists to refer to well-established knowledge, such as the molecular theory of matter, the heliocentric theory of planet revolutions, or relativity theory. Each of these scientific theories, like the theory of evolution, is not a guess or hunch as might be the case when "theory" is used in ordinary language. (Actually, scientists refer to conjectures as "hypotheses.") ID is bad science or not science at all. It is not supported by experiments, observations, or results published in peer-reviewed scientific journals. I further argue that ID is bad religion, bad theology, because it implies that the designer has undesirable attributes that we don't want to predicate about God. I hope you find the points I make in this respect convincing. ID proponents argue that the theory of evolution is incompatible with religious beliefs. Curiously, they share this conviction with materialistic scientists. I argue that both—IDers as well as materialists—are wrong: science and religion are compatible because they concern different realms of knowledge.

The last point is further developed in Chapter 9. Science is a way of knowing, but it is not the only way. Common experience, imaginative literature, art, and history provide valid knowledge about the world. The significance and purpose of the world and human life, as well as matters concerning moral or religious value, transcend science. Yet these matters are important; for most of us, they are at least as important as scientific knowledge per se.

In Chapter 10, I address a historical and epistemological question. I argue, particularly in Chapter 3, that Darwin considered natural selection, not evolution as such, his theory. I also argue that the theory of natural selection, not the evidence for evolution, is Darwin's most transcending contribution to science. The question arises, why is it that Darwin has been credited by history with the theory of evolution, more so than with the theory of natural selection? I explain that this historical mishap arises from the philosophical theory, known as empiricism, which prevailed, particularly in Britain, in the nineteenth century. Gregor Mendel, the founder of genetics, has suffered the same misattribution: He is credited with the discovery of the "laws" of inheritance, seen as generalizations derived from his experiments, rather than the discovery of the fundamental components of the "theory" of biological heredity, for which he deserves to be. Readers keen on understanding how evolution and religion are compatible may skip this chapter without missing a single beat of my argument. Readers lured by the history of ideas may find my explanations interesting and, perhaps, convincing.

I hope the arguments, explanations, and facts presented in the chapters that follow will help you comprehend why Darwin's theory of natural selection is a gift, not only to science, but to religion as well.

2

INTELLIGENT DESIGN:
THE ORIGINAL VERSION

I know no better method of introducing so large a subject,
than that of comparing a single thing with a single thing: an
eye, for example, with a telescope. As far as the examination
of the instrument goes, there is precisely the same proof that
the eye was made for vision, as there is that the telescope was
made for assisting it.

William Paley, *Natural Theology*, chap. III, p. 20

The English clergyman William Paley was intensely committed
to the abolition of the slave trade and by the 1780s had become a
much sought-after public speaker against slavery. Paley was also
an influential writer of works on Christian philosophy, ethics,
and theology. *The Principles of Moral and Political Philosophy*
(1785) and *A View of the Evidence of Christianity* (1794) earned
him prestige and well-endowed ecclesiastical benefices, which
allowed him a comfortable life. In 1800, Paley gave up his
public-speaking career for reasons of health, providing him
ample time to study science, particularly biology, and to write
*Natural Theology; or, Evidences of the Existence and Attributes of
the Deity* (1802), the book by which he has become best known
to posterity and which would greatly influence Darwin. With
Natural Theology, Paley sought to update John Ray's *Wisdom
of God Manifested in the Works of the Creation* (1691), taking
advantage of one century of additional scientific knowledge.

Paley's keystone claim is that "There cannot be design without a designer; contrivance, without a contriver; order, without choice; . . . means suitable to an end, and executing their office in accomplishing that end, without the end ever having been contemplated "[1]

Natural Theology is a sustained argument for the existence of God based on the obvious design of humans and their organs, as well as the design of all sorts of organisms, considered by themselves, as well as in their relations to one another and to their environment. The argument has two parts: first, that organisms give evidence of being designed; second, that only an omnipotent God could account for the perfection, multitude, and diversity of the designs.

There are chapters dedicated to the complex design of the human eye; to the human frame, which displays a precise mechanical arrangement of bones, cartilage, and joints; to the circulation of the blood and the disposition of blood vessels; to the comparative anatomy of humans and animals; to the digestive tract, kidneys, urethras, and bladder; to the wings of birds and the fins of fish; and much more. For 352 pages, *Natural Theology* conveys Paley's expertise: extensive and accurate biological knowledge, as detailed and precise as was available in the year 1802. After detailing the precise organization and exquisite functionality of each biological entity, relationship, or process, Paley draws again and again the same conclusion, that only an omniscient and omnipotent Deity could account for these marvels of mechanical perfection, purpose, and functionality, and for the enormous diversity of inventions that they entail.

The Eye and the Telescope

Paley's first model example in *Natural Theology* is the human eye; it appears in Chapter III, "Application of the Argument." I quote him, for there is no better way to display Paley's knowledge of the anatomy of the eye or his skill of argumentation.

Early in the chapter, Paley points out that the eye and the telescope "are made upon the same principles; both being adjusted to the laws by which the transmission and refraction of rays of light are regulated."[2] Specifically, there is a precise resemblance between the lenses of a telescope and "the humors of the eye" in their figure, their position, and the ability of converging the rays of light at a precise distance from the lens—on the retina in the case of the eye.

Paley makes two remarkable observations, which enhance the complex and precise design of the eye. The first observation is that rays of light should be refracted by a more convex surface when transmitted through water than when passing out of air into the eye. Accordingly, "the eye of a fish, in that part of it called the crystalline lens, is much rounder than the eye of terrestrial animals. What plainer manifestation of design can there be than this difference? What could a mathematical instrument maker have done more to show his knowledge of [t]his principle . . . ?"[3]

The second remarkable observation made by Paley that supports his argument is dioptric distortion.

> Pencils of light, in passing through glass lenses, are separated into different colors, thereby tinging the object, especially the edges of it, as if it were viewed through a prism. To correct this inconvenience has been long a desideratum in the art. At last it came into the mind of a sagacious optician, to inquire how this matter was managed in the eye, in which there was exactly the same difficulty to contend with as in

the telescope. His observation taught him that in the eye the
evil was cured by combining lenses composed of different
substances, that is, of substances which possessed different
refracting powers.[4]

The telescope maker accordingly corrected the dioptric distortion "by imitating, in glasses made from different materials,
the effects of the different humors through which the rays of
light pass before they reach the bottom of the eye. Could this be
in the eye without purpose, which suggested to the optician the
only effectual means of attaining that purpose?"[5]

Argument Against Chance

Paley summarizes his argument by stating the complex functional anatomy of the eye: The eye consists, "first, of a series of
transparent lenses—very different, by the by, even in their substance, from the opaque materials of which the rest of the body
is, in general at least, composed."[6] Second, the eye has the retina,
which as Paley points out is the only membrane in the body
that is black, spread out behind the lenses, so as to receive the
image formed by pencils of light transmitted through them,
and "placed at the precise geometrical distance at which, and
at which alone, a distinct image could be formed, namely, at
the concourse of the refracted rays."[7] Third, he writes, the eye
possesses "a large nerve communicating between this membrane [the retina] and the brain; without which, the action
of light upon the membrane, however modified by the organ,
would be lost to the purposes of sensation."[8]

Could the eye have come about without design or preconceived purpose, as a result of chance? Paley had set the argument
against chance, in the very first paragraph of *Natural Theology*,
reasoning rhetorically by analogy:

In crossing a heath, suppose I pitched my foot against a *stone*, and were asked how the stone came to be there, I might possibly answer, that for any thing I knew to the contrary it had lain there for ever; nor would it, perhaps, be very easy to show the absurdity of this answer. But suppose I had found a *watch* upon the ground, and it should be inquired how the watch happened to be in that place, I should hardly think of the answer which I had before given, that for any thing I knew the watch might have always been there. Yet why should not this answer serve for the watch as well as for the stone; why is it not as admissible in the second case as in the first? For this reason, and for no other, namely, that when we come to inspect the watch, we perceive—what we could not discover in the stone—that its several parts are framed and put together for a purpose, *e.g.* that they are so formed and adjusted as to produce motion, and that motion so regulated as to point out the hour of the day; that if the different parts had been differently shaped from what they are, or placed after any other manner or in any other order than that in which they are placed, either no motion at all would have been carried on in the machine, or none which would have answered the use that is now served by it.[9]

In other words, the watch's mechanism is so complicated it could not have arisen by chance.

Relation or Irreducible Complexity

The strength of the argument against chance derives, Paley tells us, from what he names "relation," a notion akin to what some contemporary authors have named "irreducible complexity."[10] This is how Paley formulates the argument. "When several different parts contribute to one effect, or, which is the same thing, when an effect is produced by the joint action of differ-ent instruments, the fitness of such parts or instruments to one another for the purpose of producing, by their united action, the effect, is what I call *relation*; and wherever this is observed in the works of nature or of man, it appears to me to carry along with it decisive evidence of understanding, intention, art."[11]

The outcomes of chance do not exhibit relation among the parts or, as we might say, they do not display organized complexity. He writes that "a wen, a wart, a mole, a pimple" could come about by chance, but never an eye; "a clod, a pebble, a liquid drop might be," but never a watch or a telescope.

Paley notices the "relation" not only among the component parts of an organ, such as the eye, the kidney, or the bladder, but also among the different parts, limbs and organs that collectively make up an animal and adapt it to its distinctive way of life: "In the *swan*, the web-foot, the spoon bill, the long neck, the thick down, the graminivorous stomach, bear all a relation to one another. . . . The feet of the mole are made for digging; the neck, nose, eyes, ears, and skin, are peculiarly adapted to an underground life. [In a word,] this is what I call relation."[12]

I am filled with amazement and respect for Paley's extensive and profound biological knowledge. He discusses the fish's air bladder, the viper's fang, the heron's claw, the camel's stomach, the woodpecker's tongue, the elephant's proboscis, the bat's wing hook, the spider's web, insects' compound eyes and metamorphosis, the glowworm, univalve and bivalve mollusks, seed dispersal, and on and on, with accuracy and as much detail as known to the best biologists of his time.

The organized complexity and purposeful function reveal, in each case, an intelligent designer, and the diversity, richness, and pervasiveness of the designs show that only the omnipotent Creator could be this Intelligent Designer.

Nature's Imperfections

Paley's natural theology flounders, however, when trying to explain how the imperfections, defects, pain, and cruelty of organisms could be consistent with his notion of the Creator. Chapter XXIII of *Natural Theology* is entitled "Of the Personality of the Deity" and it would surprise many by its well-meaning, if naïve, arrogance, as Paley seems convinced that he can determine that God is a person, God's "personality," and what his attributes are.

Paley wants, first, to establish that "contrivances," such as the eye or the kidney, cannot come about by natural principles or processes, such as Newton's laws of mechanics, which explain, for example, the motions of the planets. This is how the chapter starts: "Contrivance, if established, appears to me to prove . . . the *personality* [Paley's emphasis] of the Deity, as distinguished from what is sometimes called nature, sometimes called a principle. . . . Now, that which can contrive, which can design, must be a person. These capacities constitute personality, for they imply consciousness and thought. . . . The acts of a mind prove the existence of a mind; and in whatever a mind resides, is a person. The seat of intellect is a person"[13]

Paley then proceeds to set "the natural attributes of the Deity," namely, omnipotence, omniscience, omnipresence, eternity, self-existence, necessary existence, and spirituality—all these Paley infers from the observation of natural processes! But Paley is an honest writer who knows he has to face the difficult questions of (1) organs or parts seemingly unnecessary or superfluous and (2) imperfect and dysfunctional organs.

About seemingly superfluous organs, Paley considers two possible states of affairs: "in some instances the operation, in others the use, is unknown."[14] In some cases, we are ignorant of the function of the organ, even if we know it to be necessary for survival; in other cases, we ignore whether the organ is at all necessary. Examples of the first kind include the lungs of animals, which Paley knew to be necessary for survival, although he was not "acquainted with the action of the air upon the blood, or in what manner that action is communicated by the lungs." He cites the lymphatic system as a second example of an organ that is necessary for survival, even though how it functions was unknown. Instances "may be numerous; for they will be so in proportion to our ignorance. . . . Every improvement of knowledge diminishes their number."[15]

Examples of organs that might be unnecessary for an animal's survival include the spleen, which seems not to be necessary for "it has been extracted from dogs without any sensible injury to their vital functions."[16] However, it may well be the case that the organ serves some unknown function, even though it may not be necessary for survival in the short run.

About nature's imperfections, this is Paley's general explanation: "Irregularities and imperfections are of little or no weight, . . . but they are to be taken in conjunction with the unexceptionable evidences which we possess of skill, power, and benevolence displayed in other instances."[17]

But this account is unconvincing. If functional design manifests an Intelligent Designer, why should not deficiencies indicate that the Designer is less than omniscient, or less than omnipotent? Paley cannot have it both ways. Moreover, we know that some deficiencies are not just imperfections, but are outright dysfunctional, jeopardizing the very function the organ or part is supposed to serve. In the human eye, the optic nerve forms inside the eye cavity and creates a blind spot as it crosses the retina. This defect does not occur in the eye of squids, which is otherwise as complex as the human eye. In addition to imperfections of design, oddities, seeming cruelties, and even sadism pervade the design of the living world. Carnivorous predators behave in ways that by human standards would be judged cruel; parasites seem designed with a sadistic purpose, since they exist by harming other organisms. The mating interactions between male and female in some insects, spiders, and other organisms would be judged cruel and even sadistic by human standards. (This is a matter to which I return in Chapter 8.)

Paley is honest enough to acknowledge these difficulties as he knew them, but his explanation is inconsistent with his overall argument. Even if the dysfunctions, cruelties, and sadism of the living world were rare, which they are not, they would still need to be attributed to the Designer if the Designer had designed the living world.

Predecessors and Epigones (Pre-Darwinian)

The argument from design had been already proposed by some Fathers of the Church in the early centuries of the Christian era on the basis of the overall harmony and perfection of the universe. Augustine (354–430) affirms that "The world itself, by the perfect order of its changes and motions and by the great beauty of all things visible, proclaims . . . that it has been created, and also that it could not have been made other than by a God ineffable and invisible in greatness, and . . . in beauty."[18]

In the Middle Ages, Aquinas formulated the argument from design as the fifth way to demonstrate the existence of God. Aquinas distinguished between truths, such as the Incarnation and the Trinity, that can be known only by divine revelation, and truths accessible by human reason, which include God's existence. In his *Summa Theologiae*, Aquinas advances five ways to demonstrate, by natural reason, that God exists. The fifth way derives from the orderliness and designed purposeful-ness of the universe, which evince that it has been created by a Supreme Intelligence. "Some intelligent being exists by which all natural things are directed to their end; and this being we call God."[19]

Natural theology was disfavored by the Reformation. Martin Luther and John Calvin denied that human nature, cor-rupted after the Fall, would have the power, without Revelation, to acquire knowledge of God and his attributes.

The most forceful and elaborate formulation of the argu-ment from design, before Paley, was *The Wisdom of God Mani-fested in the Works of Creation* (1691) by John Ray (1627–1705), an English clergyman and naturalist. Ray regarded as incontro-vertible evidence of God's wisdom that all components of the universe—the stars and the planets as well as all organisms—are so wisely contrived from the beginning and perfect in their operation. The "most convincing argument of the Existence of a Deity," writes Ray, "is the admirable Art and Wisdom that discovers itself in the Make of the Constitution, the Order and Disposition, the Ends and uses of all the parts and members of this stately fabric of Heaven and Earth."[20]

On the Continent, Voltaire (1694–1778), like other philoso-phers of the Enlightenment, accepted the argument from design. Voltaire asserted that in the same way as the existence of a watch proves the existence of a watchmaker, the design and purpose evident in nature prove that the universe was created by a Supreme Intelligence.[21]

William Paley was not the only proponent of the argument from design in Britain in the first half of the nineteenth century. A few years after the publication of *Natural Theology*, the eighth Earl of Bridgewater endowed the publication of treatises that would set forth "the Power, Wisdom and Goodness of God as manifested in the Creation." Eight treatises were published during 1833–1840, several of which artfully incorporate the best science of the time and had considerable influence on the public and among scientists. William Buckland, professor of geology at Oxford University, notes in *Geology and Mineralogy* (1836) the world distribution of coal and mineral ores and proceeds to point out that they had been deposited in a remote part, yet obviously with the forethought of serving the larger human populations that would come about much later. This attribution to the Creator is particularly noteworthy because Buckland in two earlier treatises, *Vindiciae Geologicae* (1820) and *Reliquiae Diluvianae* (1823), had explained sedimentation, fossil deposits, and rock formation as natural processes, without invoking the direct intervention of God.[22] Later, another geologist, Hugh Miller in *The Testimony of the Rocks* (1858), would formulate what I call the "argument from beauty," which allows that it is not only the perfection of design but also the beauty of natural structures found in rock formations and in mountains and rivers that manifests the intervention of the Creator.

One of the Bridgewater Treatises, *The Hand, Its Mechanisms and Vital Endowments as Evincing Design*, was written by Sir Charles Bell, a distinguished anatomist and surgeon, famous for his neurological discoveries, who became professor of surgery in 1836 at the University of Edinburgh. Bell follows Paley's manner of argument, examining in considerable detail the wondrously useful design of the human hand, but also the perfection of design of the forelimb used for different purposes in different animals, serving in each case the particular needs and habits of its owner: the human's arm for handling objects, the dog's leg for running, and the bird's wing for flying. "It must now be apparent," he concluded, "that nothing less than the Power, which originally created, is equal to the effecting of those changes on animals, which are to adapt them to their conditions: that their organization is predetermined, and not consequent on the conditions of the earth or the surrounding elements."[23]

Charles Darwin, while he was an undergraduate student at the University of Cambridge between 1827 and 1831, read Paley's *Natural Theology*, which was part of the University's canon for nearly half a century after Paley's death. Darwin writes in his *Autobiography* of the "much delight" and profit that he derived from reading Paley: "In order to pass the B.A. examination, it was also necessary to get up Paley's *Evidences of Christianity*, and his *Moral Philosophy*. . . . The logic of . . . his *Natural Theology* gave me as much delight as did Euclid. . . . I did not at that time trouble myself about Paley's premises; and taking these on trust, I was charmed and convinced by the long line of argumentation."[24]

Later, however, after he returned from his five-year voyage around the world in the HMS *Beagle*, Darwin would discover a scientific explanation for the design of organisms. Science, thereby, made a quantum leap. This is the subject to which I now turn.

3

DARWIN'S REVOLUTION:
DESIGN WITHOUT DESIGNER

There is grandeur in this view of life, with its several powers, having been originally breathed into a few forms or into
one; and that . . . from so simple a beginning endless forms
most beautiful and most wonderful have been, and are being
evolved.

Charles Darwin, *The Origin of Species*, p. 490

Charles Darwin (1809–1882) occupies an exalted place in the
history of Western thought, deservedly receiving credit for the
theory of evolution. In *The Origin*, published in 1859, he laid
out the evidence demonstrating the evolution of organisms.
Darwin did not use the term "evolution," which did not have
its current meaning, but referred to the evolution of organisms
by the phrase "common descent with modification" and similar
expressions.

However, Darwin accomplished something much more
important for intellectual history than demonstrating evolution. Indeed, I proffer that accumulating evidence for common
descent with diversification may very well have been a subsidiary
objective of Darwin's masterpiece. The main subject of this
chapter is the claim that Darwin's *The Origin of Species* is, first
and foremost, a sustained effort to solve Paley's problem of how
to account scientifically for the design of organisms. Darwin

27

seeks to explain the design of organisms, their complexity, diversity, and marvelous contrivances as the result of natural processes. The evidence for evolution is brought in because evolution is a necessary consequence of his theory of design.

The introduction and chapters I through VIII of *The Origin* explain how natural selection accounts for the adaptations and behaviors of organisms, their "design." The extended argument starts in Chapter I, where Darwin describes the successful selection of domestic plants and animals and, with considerable detail, the success of pigeon fanciers seeking exotic "sports." The success of plant and animal breeders manifests how much selection can accomplish by taking advantage of spontaneous variations that occur in organisms but happen to fit the breeders' objectives. A sport (mutation) that first appears in an individual can be multiplied by selective breeding, so that after a few generations that sport becomes fixed in a breed, or "race." The familiar breeds of dogs, cattle, chickens, and food plants have been obtained by this process of selection practiced by people with particular objectives.

The ensuing chapters (II–VIII) of *The Origin* extend the argument to variations propagated by natural selection for the benefit of the organisms themselves, rather than by artificial selection of traits desired by humans. As a consequence of natural selection, organisms exhibit design, that is, exhibit adaptive organs and functions. The design of organisms as they exist in nature, however, is not "intelligent design," imposed by God as a Supreme Engineer or by humans; rather, it is the result of a natural process of selection, promoting the adaptation of organisms to their environments. This is how natural selection works: Individuals that have beneficial variations, that is, variations that improve their probability of survival and reproduction, leave more descendants than individuals of the

Charles Darwin (1809–1882), circa 1854, discoverer of natural selection. (Courtesy of G. Evelyn Hutchison.)

same species that have less beneficial variations. The beneficial variations will consequently increase in frequency over the generations; less beneficial or harmful variations will be eliminated from the species. Eventually, all individuals of the species will have the beneficial features; new features will continue accumulating over eons of time.

Organisms exhibit complex design, but it is not "irreducible complexity," emerging all of a sudden in its current elaboration.

Rather, according to Darwin's theory of natural selection, the design has arisen gradually and cumulatively, step by step, promoted by the reproductive success of individuals with incrementally more complex elaborations.

If Darwin's explanation of the adaptive organization of living beings is correct, evolution necessarily follows as a consequence of organisms becoming adapted to different environments in different localities, and to the ever-changing conditions of the environment over time, and as hereditary variations become available at a particular time that improve the organisms' chances of survival and reproduction. *The Origin*'s evidence for biological evolution is central to Darwin's explanation of "design," because this explanation implies that biological evolution occurs, which Darwin therefore seeks to demonstrate in most of the remainder of the book (chapters IX–XIII).

In the concluding Chapter XIV of *The Origin*, Darwin returns to the dominant theme of adaptation and design. In an eloquent final paragraph, Darwin asserts the "grandeur" of his vision:

> It is interesting to contemplate an entangled bank, clothed with many plants of many kinds, with birds singing on the bushes, with various insects flitting about, and with worms crawling through the damp earth, and to reflect that these *elaborately constructed* forms, *so different* from each other, and dependent on each other *in so complex a manner*, have all been produced by laws acting around us. . . . Thus, from the war of nature, from famine and death, the most exalted object which we are capable of conceiving, namely, the production of the higher animals, directly follows. There is grandeur in this view of life, with its several powers, having been originally breathed into a few forms or into one; and that, whilst this planet has gone cycling on according to the

fixed law of gravity, from so simple a beginning *endless forms most beautiful and most wonderful* have been, and are being, evolved.[1]

Darwin's *The Origin of Species* addresses the same issue as Paley: how to account for the adaptive configuration of organisms and their parts, which are so obviously "designed" to fulfill certain functions. Darwin argues that hereditary adaptive variations ("variations useful in some way to each being") occasionally appear, and that these are likely to increase the reproductive chances of their carriers. The success of pigeon fanciers and animal breeders clearly shows the occasional occurrence of useful hereditary variations. In nature, over the generations, Darwin's argument continues, favorable variations will be preserved, multiplied, and conjoined; injurious ones will be eliminated. In one place, Darwin avers: "I can see no limit to this power [natural selection] in slowly and beautifully *adapting* each form to the most complex relations of life."

In his autobiography, Darwin wrote, "The old argument of design in nature, as given by Paley, which formerly seemed to me so conclusive, falls, now that the law of natural selection has been discovered. We can no longer argue that, for instance, the beautiful hinge of a bivalve shell must have been made by an intelligent being, like the hinge of a door by a man."[2]

Natural selection was proposed by Darwin primarily to account for the adaptive organization, or "design," of living beings; it is a process that preserves and promotes adaptation. Evolutionary change through time and evolutionary diversification (multiplication of species) often ensue as by-products of natural selection fostering the adaptation of organisms to their milieu. Evolutionary change is not directly promoted by natural selection, however, and therefore it is not its necessary consequence. Indeed, some species may remain unchanged for long periods of time. Nautilus, Lingula, and other so-called "living fossils," for example, have remained unchanged in their appearance for millions of years (see below).

Evolutionary Theories

All human cultures have advanced explanations for the origin of the world and of human beings and other creatures. Traditional Judaism and Christianity explain the origin of living beings and their adaptations to life in their environments—wings, gills, hands, flowers—as the handiwork of an omniscient God. Among the early Fathers of the Church, Gregory of Nyssa (335–394) and Augustine (354–430) maintained that not all of creation, all species of plants and animals, were initially created by God; rather, some had evolved in historical times from God's creations.

According to Gregory of Nyssa, the world has come about in two successive stages. The first stage, the creative step, is instantaneous; the second stage, the formative step, is gradual and develops through time. According to Augustine, many plant and animal species were not directly created by God, but only indirectly, in their potentiality (in their *rationes seminales*), so that they would come about by natural processes, later in the world. Gregory's and Augustine's motivation was not scientific but theological. For example, Augustine was concerned that it would have been impossible to hold representatives of all animal species in a single vessel, such as Noah's Ark; some species must have come into existence only after the Flood.

The notion that organisms may change by natural processes was not investigated as a biological subject by Christian theologians of the Middle Ages, but it was, usually incidentally, considered as a possibility by many, including Albertus Magnus

(1200–1280) and his student Thomas Aquinas (1224–1274). Aquinas concluded, after consideration of the arguments, that the development of living creatures, such as maggots and flies, from nonliving matter, such as decaying meat, was not incompatible with Christian faith or philosophy, but he left it to others (to scientists, in current parlance) to determine whether this actually happened.

The issue whether living organisms could spontaneously arise from dead matter was not settled until four centuries later by the Italian Francesco Redi (1626–1697), one of the first scientists to conduct biological experiments with proper controls. Redi set up flasks with various kinds of fresh meat; some were sealed, others covered with gauze so that air but not flies could enter, and others left uncovered. The meat putrefied in all flasks, but maggots appeared only in the uncovered flasks which flies had entered freely. Redi was a poet as well as a physician, chiefly known for his *Bacco in Toscana* (1685, Baccus in Tuscany).

The cause of putrefaction would be discovered two centuries later by Darwin's younger contemporary, the French chemist Louis Pasteur (1822–1895), one of the greatest scientists of all time. Pasteur demonstrated that fermentation and putrefaction were caused by minute organisms that could be destroyed by heat. Food decomposes when placed in contact with germs present in the air. The germs do not arise spontaneously within the food. We owe to Pasteur the process of pasteurization, the destruction by heat of microorganisms in milk, wine, and beer, which can thus be preserved if kept out of contact with the microorganisms in the air. Pasteur also demonstrated that cholera and rabies are caused by microorganisms and he invented vaccination, treatment with attenuated (or killed) infective agents that would stimulate the immune system of animals and humans, thus protecting them against infection.

The first broad theory of evolution was proposed by the French naturalist Jean-Baptiste de Monet, Chevalier de Lamarck (1744–1829). In his *Philosophie zoologique* (1809, Zoological Philosophy), Lamarck held the enlightened view, shared by the intellectuals of his age, that living organisms represent a progression, with humans as the highest form. Lamarck's theory of evolution asserts that organisms evolve through eons of time from lower to higher forms, a process still going on, always culminating in human beings. The remote ancestors of humans were worms and other inferior creatures, which gradually evolved into more and more advanced organisms, ultimately humans.

The "inheritance of acquired characters" is the theory most often associated with Lamarck's name. Yet this theory was actually a subsidiary construct of his theory of evolution: that evolution is a continuous process and that today's worms will yield humans as their remote descendants. It stated that as animals become adapted to their environments through their habits, modifications in their body plans occur by "use and disuse." Use of an organ or structure reinforces it; disuse leads to obliteration. The theory further asserted that the characteristics acquired by use and disuse, according to Lamarck, would be inherited. This assumption would later be called the inheritance of acquired characteristics (or Lamarckism). It was disproved in the twentieth century.

Lamarck's evolutionary theory was metaphysical rather than scientific. He postulated that life possesses the innate property to improve over time, so that progression from lower to higher organisms would continually occur, and always following the same path of transformation from lower organisms to increasingly higher and more complex organisms. A somewhat similar evolutionary theory was formulated one century later by another Frenchman, the philosopher Henri Bergson (1859–1940) in his *L'Evolution créatrice* (1907, Creative Evolution).

Erasmus Darwin (1731–1802), a physician and poet, and the grandfather of Charles Darwin, proposed, in poetic rather than scientific language, a theory of the transmutation of life forms through eons of time (*Zoonomia, or the Laws of Organic Life*, 1794–1796). More significant for Charles Darwin was the influence of his older contemporary and friend, the eminent geologist Sir Charles Lyell (1797–1875). In his *Principles of Geology* (1830–1833), Lyell proposed that Earth's physical features were the outcome of major geological processes acting over immense periods of time, incomparably greater than the few thousand years since Creation that was commonly believed at the time.

Charles Darwin and the Voyage of the *Beagle*

Charles Darwin was the son and grandson of physicians. He enrolled as a medical student at the University of Edinburgh. After two years, however, he left Edinburgh and moved to the University of Cambridge to pursue studies that would prepare him to become a clergyman. Darwin was not an exceptional student, but he was deeply interested in natural history. On December 27, 1831, a few months after his graduation from Cambridge, he sailed as a naturalist aboard the HMS *Beagle* on a round-the-world trip that lasted until October 1836. On that voyage, Darwin often disembarked for extended trips ashore to collect natural specimens. The discovery of fossil bones from large extinct mammals in Argentina and the observation of numerous species of finches in the Galápagos Islands were among the events credited with stimulating Darwin's interest in how species originate.

The voyage of the *Beagle* (1831–1836).

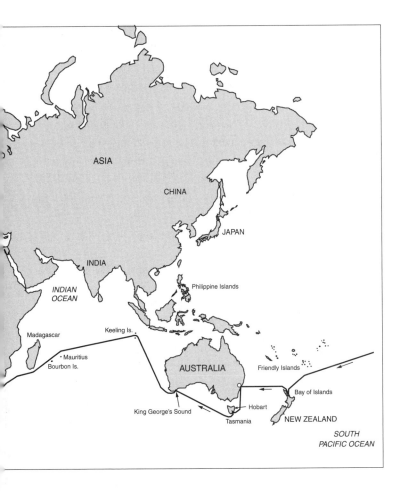

His Galápagos Islands observations may have been the most influential on Darwin's thinking. The islands, which lie on the equator 600 miles off the west coast of South America, had been named Galápagos (the Spanish word for tortoises) by the Spanish explorers who discovered them, because of the abundance of giant tortoises, different ones on different islands and all of them different from those known anywhere else in the world. The Galápagos tortoises sluggishly clanked their way around, feeding on the vegetation and seeking the few pools of fresh water. They would have been vulnerable to predators, had there been any on the islands.

Darwin also found large lizards that feed, unlike any others of their kind, on seaweed, and mockingbirds, quite different from those found on the South American mainland. His observations of several kinds of finches that varied from island to island in their features, notably their distinctive beaks that were adapted to disparate feeding habits—crushing nuts, probing for insects, grasping worms—are now part of the canon of science history.

In addition to *The Origin of Species* (1859), Darwin published many other books, notably *The Descent of Man, and Selection in Relation to Sex* (1871), which extends the theory of natural selection to human evolution.

Copernicus and Darwin

There is a version of the history of the ideas that sees a parallel between the Copernican and the Darwinian revolutions. In this view, the Copernican Revolution consisted of displacing the earth from its previously accepted locus as the center of the universe, moving it to a subordinate place as just one more planet revolving around the sun. In congruous manner,

Testudo microphyes, Isabela I.

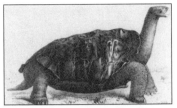

Testudo abingdonii, Pinta I.

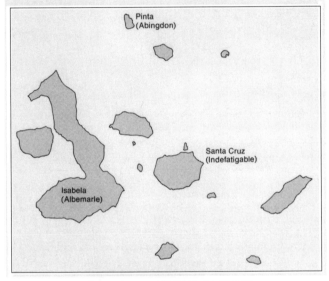

Testudo ephippium, Santa Cruz I.

The Galápagos Islands, with drawings of three tortoises found in different islands.

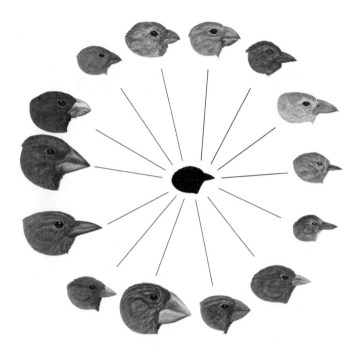

Darwin's finches. Fourteen species from the Galápagos Islands that evolved from a common ancestor. Different species feed on different foods and have evolved beaks adapted to their feeding habits. (Adapted from "Evolution, The Theory of," courtesy of Encyclopaedia Britannica, Inc.)

the Darwinian Revolution is viewed as consisting of the displacement of humans from their exalted position as the center of life on earth, with all other species created for the service of humankind. According to this version of intellectual history, Copernicus had accomplished his revolution with the heliocentric theory of the solar system. Darwin's achievement emerged from his theory of organic evolution.[3]

This version of the two revolutions is inadequate: What it says is true, but it misses what is most important about these two intellectual revolutions, namely that they ushered in the beginning of science in the modern sense of the word. These two revolutions may jointly be seen as the one Scientific Revolution, with two stages, the Copernican and the Darwinian.

The Copernican Revolution was launched with the publication in 1543, the year of Nicolaus Copernicus' death, of his *De revolutionibus orbium celestium* (On the Revolutions of the Celestial Spheres), and bloomed with the publication in 1687 of Isaac Newton's *Philosophiae naturalis principia mathematica* (The Mathematical Principles of Natural Philosophy). The discoveries by Copernicus, Kepler, Galileo, Newton, and others, in the sixteenth and seventeenth centuries, had gradually ushered in a conception of the universe as matter in motion governed by natural laws. It was shown that Earth is not the center of the universe, but a small planet rotating around an average star; that the universe is immense in space and in time; and that the motions of the planets around the sun can be explained by the same simple laws that account for the motion of physical objects on our planet.[4]

These and other discoveries greatly expanded human knowledge. The conceptual revolution they brought about was more fundamental yet: a commitment to the postulate that the universe obeys immanent laws that account for natural phenomena. The workings of the universe were brought into the realm of science: explanation through natural laws. All physical phenomena could be accounted for as long as the causes were adequately known.

The advances of physical science brought about by the Copernican Revolution had driven mankind's conception of the universe to a split-personality state of affairs, which persisted well into the mid-nineteenth century. Scientific explanations, derived from natural laws, dominated the world of nonliving matter, on the Earth as well as in the heavens. Supernatural explanations, such as Paley's explanation of design, which depended on the unfathomable deeds of the Creator, accounted for the origin and configuration of living creatures—the most diversified, complex, and interesting realities of the world.

It was Darwin's genius to resolve this conceptual schizo-phrenia. Darwin completed the Copernican Revolution by drawing out for biology the notion of nature as a lawful system of matter in motion that human reason can explain without recourse to supernatural agencies. The conundrum faced by Darwin can hardly be overestimated. The strength of the argu-ment from design to demonstrate the role of the Creator had been forcefully set forth by Paley. Wherever there is function or design, we look for its author. It was Darwin's greatest accom-plishment to show that the complex organization and func-tionality of living beings can be explained as the result of a natural process—natural selection—without any need to resort to a Creator or other external agent. The origin and adaptation of organisms in their profusion and wondrous variations were thus brought into the realm of science.

Darwin accepted that organisms are "designed" for certain purposes, that is, they are functionally organized. Organisms are adapted to certain ways of life and their parts are adapted to perform certain functions. Fish are adapted to live in water, kidneys are designed to regulate the composition of blood, the human hand is made for grasping. But Darwin went on to provide a natural explanation of the design. The seemingly pur-poseful aspects of living beings could now be explained, like the phenomena of the inanimate world, by the methods of science, as the result of natural laws manifested in natural processes.

Darwin's "Theory"

The conclusion that Darwin considered natural selection (rather than his demonstration of evolution) his most important discovery, emerges from consideration of his life and works. Darwin himself treasured natural selection as his greatest discovery and designated it as "my theory," a designation he never used when referring to the evolution of organisms. The discovery of natural selection, Darwin's awareness that it was a greatly significant discovery because it was science's answer to Paley's argument from design, and Darwin's designation of natural selection as "my theory" can be traced in Darwin's "Red Notebook" and "Transmutation Notebooks B to E," which he started in March 1837, not long after returning (on October 2, 1836) from his voyage on the *Beagle*, and completed in late 1839.[5]

The evolution of organisms was commonly accepted by naturalists in the middle decades of the nineteenth century, and the distribution of exotic species in South America, in the Galápagos Islands, and elsewhere, and the discovery of fossil remains of long-extinguished animals, confirmed the reality of evolution in Darwin's mind. The intellectual challenge was to explain the origin of distinct species of organisms, how new ones adapted to their environments, that "mystery of mysteries," as it had been labeled by Darwin's older contemporary, the prominent scientist and philosopher Sir John Herschel (1792–1871).

Early in the Notebooks of 1837 to 1839, Darwin registers his discovery of natural selection and repeatedly refers to it as "my theory." From then until his death in 1882, Darwin's life would be dedicated to substantiating natural selection and its companion postulates, mainly the pervasiveness of hereditary variation and the enormous fertility of organisms, which much surpassed the capacity of available resources. Natural selection became for Darwin "a theory by which to work." He relentlessly pursued observations and performed experiments in order to test the theory and resolve presumptive objections.

Alfred Russel Wallace (1823–1913) is famously given credit for discovering, independently of Darwin, natural selection as the process accounting for the evolution of species. On June 18, 1858, Darwin wrote to Charles Lyell that he had received by mail a short essay from Wallace such that "if Wallace had my [manuscript] sketch written in [1844] he could not have made a better abstract." Darwin was thunderstruck.

Darwin and Wallace started occasional correspondence in late 1855 at the time Wallace was in the Malay archipelago collecting biological specimens. In his letters, Darwin would offer sympathy and encouragement to the occasionally dispirited Wallace for his "laborious undertaking." In 1858, Wallace came upon the idea of natural selection as the explanation for evolutionary change and he wanted to know Darwin's opinion about this hypothesis, since Wallace, as well as many others, knew that Darwin had been working on the subject for years, had shared his ideas with other scientists, and was considered by them as the eminent expert on issues concerning biological evolution.

Darwin was uncertain how to proceed about Wallace's letter. He wanted to credit Wallace's discovery of natural selection, but he did not want altogether to give up his own earlier independent discovery. Eventually, Sir Charles Lyell and Joseph Hooker proposed, with Darwin's consent, that Wallace's letter and two of Darwin's earlier writings would be presented at a meeting of the Linnean Society of London. On July 1, 1858, three papers were read by the society's undersecretary, George Busk, in the order of their date of composition: Darwin's abbreviated abstract of his 230-page essay from 1844; an "abstract of abstract" that Darwin had written to the American botanist Asa Gray on September 5,

1857; and Wallace's essay, "On the Tendency of Varieties to Depart Indefinitely from Original Type; Instability of Varieties Supposed to Prove the Permanent Distinctness of Species."[6] The meeting was attended by some thirty people, who did not include Darwin or Wallace. The papers generated little response and virtually no discussion, their significance apparently lost to those in attendance. Nor was it noticed by the president of the Linnean Society, Thomas Bell, who, in his annual address the following May, blandly stated that the past year had not been enlivened by "any of those striking discoveries which at once revolutionize" a branch of science.

Wallace's independent discovery of natural selection is remarkable. Wallace, however, was not interested in explaining design, but rather in accounting for the evolution of species, as indicated in his paper's title: "On the Tendency of Varieties to Depart Indefinitely from the Original Type." Wallace thought that evolution proceeds indefinitely and is progressive.[7] Darwin, on the contrary, did not accept that evolution would necessarily represent progress or advancement, nor did he believe that evolution would always result in morphological change over time; rather, he knew of the existence of "living fossils," organisms that had remained unchanged for millions of years. For example, "some of the most ancient Silurian animals, as the Nautilus, Lingula, etc., do not differ much from living species."[8]

In 1858, Darwin was at work on a multivolume treatise, intended to be titled "On Natural Selection." Wallace's paper stimulated Darwin to write *The Origin*, which would be published the following year. Darwin intended this as an abbreviated version of the much longer book he had intended to write. As noted earlier, Darwin's focus, in *The Origin* as elsewhere, was the explanation of design, with evolution playing the subsidiary role of supporting evidence.

The Darwinian Aftermath

The publication of *The Origin* produced considerable public excitement. Scientists, politicians, clergymen, and notables of all kinds read and discussed the book, defending or deriding Darwin's ideas. The most visible actor in the controversies immediately following publication was the English biologist T. H. Huxley, who later became known as "Darwin's bulldog," and who defended the theory of evolution with articulate and sometimes mordant words, on public occasions as well as in numerous writings.

A younger English contemporary of Darwin, with considerable influence over the public during the latter part of the nineteenth and in the early twentieth century, was Herbert Spencer. A philosopher rather than a biologist, he became an energetic proponent of evolutionary ideas, popularized a number of slogans, such as "survival of the fittest" (which was taken up by Darwin in later editions of *The Origin*), and engaged in social and metaphysical speculations. Unfortunately, his mistaken ideas considerably damaged proper understanding and acceptance of the theory of evolution by natural selection. Darwin wrote of Spencer's speculations that "his deductive manner of treating any subject is wholly opposed to my frame of mind. . . . His fundamental generalizations are of such a nature that they do not seem to me to be of any strictly scientific use." Most pernicious was the crude extension by Spencer and others of the notion of the "struggle for existence" to human economic and social life that became known as Social Darwinism.

Darwin's theory of evolution was strenuously debated in the latter part of the nineteenth century and beyond. In the first three decades of the twentieth century, the controversy centered on the importance of genetic mutations relative to natural selection. By the middle of the twentieth century, theoretical advances and the accumulation of experimental evidence resulted in the formulation of the modern theory of evolution, which is universally accepted by biologists. Knowledge of evolutionary processes has continued to advance, and at an accelerated rate, into the present, often promoted by discoveries in other disciplines, such as molecular biology.

I turn to the evidence for evolution later, starting in Chapter 5, but first, in Chapter 4, I introduce natural selection, the central construct of Darwin's theory.

4

Natural Selection

I can see no limit to this power [natural selection], in slowly and beautifully adapting each form to the most complex relations of life.

Charles Darwin, *The Origin of Species*, p. 469

Natural selection was proposed by Darwin primarily to account for the adaptive organization of living beings; it is a process that maintains and promotes adaptation. Evolutionary change through time and evolutionary diversification (multiplication of species) are not directly promoted by natural selection, but they often ensue as by-products of natural selection as it fosters adaptation to different environments.

A summary of Darwin's theory of natural selection, in his own words, is the following:

Can it, then, be thought improbable, seeing that variations useful to man have undoubtedly occurred, that other variations useful in some way to each being in the great and complex battle of life, should sometimes occur in the course of thousands of generations? If such do occur, can we doubt (remembering that more individuals are born than can possibly survive) that individuals having any advantage, however slight, over others, would have the best chance

of surviving and of procreating their kind? On the other
hand, we may feel sure that any variation in the least degree
injurious would be rigidly destroyed. This preservation of
favorable variations and the rejection of injurious variations,
I call Natural Selection.[1]

What Darwin is saying is this. The experience of agriculture
has taught us that animals and plants from time to time show
new variants in their traits, so that farmers can select desirable
features, say, corn with larger kernels or cows that produce
more milk. These variants are heritable, that is, are transmitted
to the offspring. Surely, says Darwin, variants that are beneficial
to the organisms themselves must also occur from time to time,
such as increased running speed in a cheetah or better dispersal
of seed in an oak. These variants are "useful" to the organisms
precisely because they increase their chances for survival and
procreation. This in turn means that these advantageous variations will be multiplied over the generations at the expense of
less advantageous variants. This is the process known as "natural selection," and evolution occurs as a consequence. The key
point is that natural selection—the "preservation of favorable
variations and the rejection of injurious variations," in Darwin's
words—accounts for the "design" of organisms, why they are
well constructed so that they can function in the environments
where they live. A fleeting cheetah captures more prey; a tree
with more leaves captures sunlight more effectively.

The Concept of Natural Selection

Darwin dedicated much of *The Origin* to explaining how natural selection works. Much more has been learned by scientists
in the past half century. If a short definition that catches the
core of the process is desired, we can say that natural selection is

"the differential reproduction of hereditary variations," which is how textbooks often define it in short. That is saying simply that useful variants multiply more effectively over the generations than less useful (or harmful) variants. Thus a cheetah able to run faster will catch more prey, and therefore live longer and leave more offspring than a slower cheetah. So, a hereditary variant that increases fleetness will increase in frequency over the generations and eventually replace the slower variant.

It is obvious that the brief definition given in the preceding paragraph does not provide a satisfactory understanding of the process of natural selection and how it accounts for the evolution of organisms and their design. Just as we would not learn much about the Earth by defining it as "the third planet that revolves around the sun." We can increase our understanding by extending the definition as follows: "Natural selection is the differential reproduction of alternative variations, determined by the fact that some variations are beneficial because they increase the probability that the organisms having them will live longer or be more fertile than organisms having alternative variations." To the previous very short definition, this one adds the reason why differential reproduction occurs, namely because some variants are more useful than others. It also specifies the two main components of reproduction: survival and fertility. We might want to make the definition more informative yet by referring to the outcome of the process, adding to the previous definition: "Over the generations, beneficial variations will be preserved and multiplied; injurious or less beneficial variations will be eliminated." We could also refer in the definition to the long-term consequences of the process: "Over long periods of time, natural selection usually changes the makeup and functioning of organisms and causes their diversification (i.e., multiplication of species) as they adapt to different environments."

However, it is not possible to formulate a *satisfactory* definition of natural selection. The point that I wish to make with the preceding paragraphs is that a simple definition may help us to focus on some important aspect of natural selection, but natural selection—like any other complex or interesting process in the world—cannot be satisfactorily embraced in just a few words. We might want to define the molecular theory of matter by saying that it states that "every physical entity, whether solid, liquid, or gas is made up of molecules"; and we might define plate tectonics as "the motion of the continental plates around the Earth." In both cases, there is much more to know about the two theories than what is stated in the brief definitions; additional knowledge is conveyed in numerous scientific articles and books about the molecular composition of matter and about plate tectonics.

Similarly, there are many books and innumerable scientific papers that expound on the complexities of the process of natural selection: books and papers in which appropriate mathematical models and equations are developed that account for the process of natural selection, and other papers and books that report laboratory experiments or investigations of natural selection in nature. Darwin dedicated much of *The Origin* to demonstrating and explaining natural selection. Moreover, Darwin wrote several other books further describing how natural selection works, books dedicated, for example, to the evolution of orchids, barnacles, earthworms, primates, and humans. The rest of this chapter focuses on some important features of the process of natural selection but, of course, without being exhaustive, which would take a much, much longer book.

Darwin's Monk

Biological evolution is the process of change and diversification of living things over time, and it affects all aspects of their lives—morphology (form and structure), physiology (function), behavior, and ecology (interaction with the environment). Underlying these changes are changes in the hereditary materials. Hence, in genetic terms, evolution consists of changes in the organism's hereditary makeup.

Evolution can be seen as a two-step process. First, hereditary variation arises by "mutation"; second, selection occurs by which useful variants increase in frequency and those less useful or injurious are eliminated over the generations. As explained above, the variants that arise by mutation are not transmitted equally from one generation to another.[2] Some become more frequent because they increase the ability of the organism to survive and/or produce more offspring. What we need now is to explain the first step of the process: how mutations occur and how they are inherited.

The most serious difficulty facing Darwin's explanation of design was precisely that it was not known how the first step of the process occurred. In fact, the understanding of how biological heredity occurs was completely wrong. Contemporary theories of "blending inheritance" proposed that offspring merely struck an average between the characteristics of their parents, just as if we mix red and white liquid paint, we get pink. However, as Darwin became aware, blending inheritance could not account for the conservation of variations, because differences between variant offspring would be halved each generation, rapidly reducing the advantage of any mutation as it became averaged again and again over the generations; it would gradually lose its distinctness and therefore any possible advantages it might originally have had over preexisting characteristics. (This issue is discussed in more detail in the last chapter of this book, which is quite technical in some respects and may not be of interest to all readers. So, I'll summarize the matter here.)

The missing link in Darwin's argument was provided by Mendelian genetics, the work of an obscure Augustinian monk, Gregor Mendel. About the time *The Origin* was published, Mendel started a long series of experiments with peas in the garden of his monastery in Brünn, Austria-Hungary (now Brno, Czech Republic). These experiments and the analysis of their results, which are by any standard a masterly example of the scientific method, enabled Mendel to formulate the fundamental principles of the theory of heredity that is still in use today. This theory accounts for biological inheritance through particulate factors (now known as "genes"), which occur in pairs and are inherited one from each parent. The key point is that the two genes for each trait do not mix or blend but segregate in the formation of the sex cells, or gametes. For example, one gene causes the pea plant to be tall; the alternative gene causes the plant to be short. When a tall plant and a short plant are crossed, the hybrid inherits one gene from each parent. In hybrids generally, and not only in peas, the hybrid may be like one parent, like the other, or intermediate between both. The key discovery of Mendel is that the genes retain their distinctness in the hybrid and are so transmitted by the hybrid to its progeny. It is as if when you mix red and white paint, you may get pink, but when you pick up bits of the pink paint, some are red and some are white. If you mix the bits that are red, you end with red paint again. This is quite counterintuitive. No wonder it took many years and the genius of Mendel to discover it.

Mendel's discoveries were unknown to Darwin, however, and, indeed, they did not become generally known until 1900, when they were simultaneously rediscovered by three European scientists.[3]

Hereditary variations, favorable or not to the organisms, arise by a process known as mutation, which changes a gene into another. For example, a gene that causes a plant to be short mutates into a gene that causes a plant to be tall. The details of the mutation process were gradually elucidated in the twentieth century. With respect to evolution, what is significant is that unfavorable mutations are eliminated by natural selection because their carriers leave no descendants or leave fewer than those carrying alternative favorable mutations. Favorable mutations accumulate over the generations because the organisms in which they occur leave a greater number of descendants than those with unfavorable mutations. The process of natural selection continues indefinitely because the environments that organisms inhabit are forever changing. Environments change physically—in their climate, configuration, and so on—but also biologically, because the predators, parasites, competitors, and food sources with which an organism interacts are themselves evolving. Different mutations are favored in different environments, or "habitats"; as the habitats change (or organisms colonize new ones), the organisms will evolve. Mutation is a random process. Mutations arise without regard to their effects on the organisms' ability to survive and reproduce.

If mutation were the only process affecting evolutionary change, the organization of living things would gradually disintegrate. The effects of mutation alone would be analogous to those of a mechanic who changed parts in an automobile engine at random, with no regard for the role of the parts in the engine. Natural selection keeps the disorganizing effects of mutation in check because it multiplies beneficial mutations and eliminates harmful ones.

Natural selection accounts not only for the preservation and improvement of the organization of living beings but also for their diversity. In different localities or in different circumstances, natural selection favors different traits, precisely those that make the organisms well adapted to their particular circumstances and ways of life.

Mutations and DNA

Life on Earth is thought to have originated about 3.5 billion years ago, although some experts place it later. The primordial organisms were very small and relatively simple, yet all living things have evolved from these lowly beginnings. At present, there are more than 2 million known species, which are widely diverse in size, shape, and way of life. This diversity is grounded on molecules of DNA (deoxyribonucleic acid) that reside in each and every cell of an organism.

DNA is a double-helix molecule that carries information in the long sequence of its four components ("nucleotides"), which scientists represent by the letters A, C, G, and T (which stand for adenine, cytosine, guanine, and thymine). The genetic information is contained in sequences of these nucleotides, in a manner analogous to the way semantic information is conveyed by sequences of letters of the English alphabet. The amount of genetic information in organisms is enormous because the total length of the DNA molecules of an organism is huge. For example, the human genome—that is, the DNA that each human inherits from each parent—is 3 billion letters long. Printing one human genome would require 1,000 books, each 1,000 pages long, with 3,000 letters (about 500 words) per page. Scientists do not print full genomes of humans or other organisms; rather, the DNA information is stored electronically in computers.

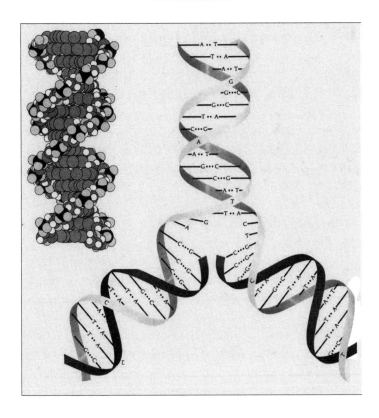

Left. The double helix of DNA consists of two strands twisting around each other. (DNA is made up of five kinds of atoms represented as circles: hydrogen, oxygen, carbon, nitrogen, and phosphorus.) *Right.* During replication, the two strands unwind and each one serves as a template for the synthesis of a complementary strand, so that the two daughter double helices are identical to each other and to the original molecule. There are four kinds of components (nucleotides), represented by A, C, G, and T, the letters of the genetic alphabet. In the double helix, A pairs only with T, and C with G. The genetic information is embodied in the sequence of letters along the DNA—3 billion of them in each human genome.

The information encoded in the nucleotide sequence of DNA is, as a rule, faithfully reproduced during replication, so that each replication results in two DNA molecules that are identical to each other and to the parent molecule. Heredity is not a perfectly conservative process, however. Occasionally mutations occur in the DNA molecule during replication, so that daughter cells differ from the parent cells in the sequence or in the amount of DNA. Mutations often involve one single letter (nucleotide), but occasionally mutations may encompass several or many letters. A mutation first appears in the DNA in a single cell of an organism, and that new, changed DNA is passed on to all cells descended from the first. The mutations that count in evolution are those that occur in the sex cells (eggs and sperm), or in cells from which the sex cells derive, because these are the cells that produce the next generation.

Mutation rates have been measured in a great variety of organisms, mostly for mutants that exhibit conspicuous effects. In humans and other multicellular organisms, the rate typically ranges around 1 mutation per 1,000,000 sex cells.[4] Although mutation rates are low, new mutants appear continuously in nature because there are many individuals in every species and many genes in every individual. The human population consists of more than 6 billion people. If any given mutation occurs once for each million people, living humans would collectively carry 6,000 copies of every possible mutation.

The process of mutation provides each generation with many new genetic variations, in addition to those carried over from previous generations. Thus, it is not surprising to see that, when new environmental challenges arise, species are able to adapt to them. More than 100 insect species, for example, have developed resistance to the pesticide DDT in parts of the world where spraying has been intense. Although these animals had never before encountered this synthetic compound, mutations gave them some advantage that allowed them to survive in its presence. That "adaptation" was rapidly multiplied by natural selection.

The resistance of disease-causing bacteria and parasites to antibiotics and other drugs is a consequence of the same process. When an individual receives an antibiotic that specifically kills the bacteria causing a disease—say, tuberculosis—the immense majority of the bacteria die, but one in several million may have a mutation that provides resistance to the antibiotic. These resistant bacteria will survive and multiply, and that antibiotic will no longer cure the disease. This is why modern medicine treats bacterial diseases with cocktails of antibiotics. If the incidence of a mutation conferring resistance for a given antibiotic is one in a million, the incidence of one bacterium carrying three mutations, each conferring resistance to one of three antibiotics, is one in a quintillion (one in a million million million); it is not likely, if not altogether impossible, for such bacteria to exist in any infected individual.

Natural Selection as a Creative Process

Natural selection is sometimes perceived as a "purifying" process, the elimination of harmful mutations. But natural selection is much more than a purely negative process, for it is able to generate novelty by increasing the probability of otherwise extremely improbable genetic combinations. Natural selection is thus a creative process. Although it does not "create" the component entities upon which it operates (genetic mutations), it does yield adaptive combinations that could not have existed otherwise.

The combination of genetic units that carry the hereditary information responsible for the formation of the vertebrate eye could never have been produced purely by a random process, not even if we allow for the 3-billion-plus years during which life has existed on Earth. This is the argument advanced by proponents of intelligent design. However, evolution is not a process governed by random events. The complicated anatomy of the eye, like the exact functioning of the kidney, are the result of a nonrandom process—natural selection.

How natural selection can generate novelty in the form of accumulated hereditary information may be illustrated by the following example of an experiment made with *Escherichia coli*, single-celled bacteria that occur in the colon of humans and other mammals. Some strains of *E. coli*, in order to reproduce in a culture (a small test-tube with a water solution of sugar), require that a certain substance, the amino acid histidine, be provided with the sugar. When a few such bacteria are added to a small test-tube with a culture solution that includes histidine, they multiply rapidly and produce between 20 and 30 billion bacteria in one or two days. If a drop of the antibiotic streptomycin is added to the culture, most bacteria will die, but after a day or two the culture will again teem with billions of bacteria. How come?

Spontaneous genetic mutations causing resistance to streptomycin occur in normal (i.e., nonresistant) bacteria randomly, at rates on the order of 1 in 100 million bacterial cells. In a bacterial culture with 20 to 30 billion bacteria, we expect between 200 and 300 bacteria to be resistant. When streptomycin is added to the culture, only the resistant cells survive. The 200 or 300 surviving bacteria will start reproducing, however, and allowing one or two days for the necessary number of cell divisions, twenty or so billion bacteria are produced, all resistant to streptomycin.

Consider now a second step in this experiment. The streptomycin-resistant cells are transferred to a culture with streptomycin but without histidine (the amino acid that they require in order to grow and reproduce). Most of the bacteria will fail to reproduce and will die; yet, after a day or two, the culture will be teeming with billions of bacteria. This is because among cells that need the amino acid histidine to grow, mutants able to reproduce in the absence of histidine arise spontaneously at rates of about 4 in 100 million bacteria. If the culture has 20 to 30 billion bacteria, about 1,000 bacteria will survive in the absence of histidine and will start reproducing until the available medium is saturated with them.

Thus, natural selection has produced, in two steps, bacterial cells resistant to streptomycin and not requiring histidine for growth. The probability of these two mutations happening in the same bacterium is about 4 in 10 million billion cells. An event of such low probability is unlikely to occur even in a large laboratory culture of bacterial cells. Yet natural selection commonly results in cells possessing both properties. A "complex" trait made up of two components has come about by natural processes. It can readily be understood that the example can be extended to three, four, and more component steps. At the end of the long process of evolution, we have organisms each exhibiting features "designed" for its survival in its habitat.

A Monkey's Tale

Critics have sometimes alleged, against Darwin's theory of evolution, examples or arguments that, they claim, show that random processes cannot yield meaningful, organized outcomes. It is pointed out, for example, that monkeys, even a large number of them, randomly striking letters on a typewriter, would never write *The Origin of Species*, even if we allow for millions of years and many generations of monkeys pounding at typewriters. This argument is cogent against any process that is random. But natural selection is not a random process. It is a process that promotes adaptation by selecting combinations that "make sense," that is, that are useful to the organisms. Consider the following modification of the monkey example.

A process exists by which meaningful words are chosen whenever they appear on the typewriter; words such as "the," "sun," "also," "rises," etc. These simple combinations of a few letters will occasionally arise. Assume further that any arising words are transferred to the keys of another typewriter. The random strikes of the monkeys on the keys of this second-level typewriter will, on occasion, yield word combinations, such as "the sun also rises." Whenever meaningful combinations of words (i.e., sentences) occur, they are transferred to the keys of a third-level typewriter, on which meaningful paragraphs that arise are selected and incorporated into the keys of a higher-order typewriter, and so on. It is clear that pages and even chapters "making sense" would eventually be produced. Yet, the end product would not be an "irreducibly complex" text. In nature, it is the process of natural selection that "picks up" the combinations that "make sense," as in the bacterial example.

I need not carry the monkey analogy far, since it is far from satisfactory. The point that I want to emphasize, for those who argue that design and adaptation to the environment cannot come about by random processes, is that evolution is not the outcome of random processes. There is a "selecting" process that picks up adaptive combinations because these reproduce more effectively and thus come to prevail in populations. Simple adaptive combinations constitute, in turn, new levels of organization upon which the mutation (random) plus selection (nonrandom or directional) processes again operate. The organizational complexity of animals and plants has arisen as a consequence of natural selection acting one step at a time, over eons of time.

Several hundred million generations separate modern animals from the early animals of the Cambrian geological period (542 million years ago). The number of mutations that can be tested, and those eventually selected, in millions of individual animals over millions of generations is difficult for a human mind to fathom, but we can readily understand that the accumulation of millions of small, functionally advantageous changes could yield remarkably complex and adaptive organs, such as the eye.

Another critical point is that evolution by natural selection is an incremental process, operating over time and yielding organisms better able to survive and reproduce than others, which typically differ from one another at any one time only in small ways, for example, the difference between having or lacking an enzyme able to synthesize the amino acid histidine. Numerous adaptations are known that involve one or only a few genes, as in the bacterial example. An example occurs in some pocket mice (*Chaetodipus intermedius*) that live in rocky outcrops in Arizona. Light, sandy-colored mice are found in light-colored habitats, whereas dark (melanic) mice prevail in dark rocks formed from ancient flows of basaltic lava. The match between background and fur color protects the mice from avian and mammal predators that hunt guided largely by vision. Mutations in one single gene (coding for the melanocortin-1-receptor, represented as *MC1R*) account for the difference between light and dark peltage.[5]

Adaptations that involve complex structures, functions, or behaviors typically involve numerous genes. Many familiar mammals, but not marsupials, have a placenta. Marsupials include the familiar kangaroo and many other mammals native primarily to Australia and South America. Dogs, cats, mice, donkeys, and primates are placental. The placenta makes it possible to extend the time the developing embryo is kept inside the mother and thus make the newborn better prepared for independent survival. However, the placenta requires complex adaptations, such as the suppression of harmful immune interactions between mother and embryo, delivery of suitable nutrients and oxygen to the embryo, and the disposal of embryonic wastes. The mammalian placenta evolved more than 100 million years ago and proved a successful adaptation, leading to the extinction of most marsupial species in the continents of the Old World and North America. The placenta also has evolved in some fish groups, such as *Poeciliopsis*. In some species, females supply the yolk in the egg, which furnishes nutrients to the developing embryo (as in chicken), but do not directly contribute nutrients to the embryo. Other species, however, have evolved a placenta through which the mother provides additional nutrients to the developing embryo. The reconstruction of the evolutionary history of *Poeciliopsis* species, by means of molecular biology, has shown that the placenta evolved independently three times in this fish group and that the required complex adaptations accumulated in less than 750,000 years.[6]

It is worth pointing out that increased complexity is not a necessary consequence of natural selection, but emerges occasionally. Occasionally, a mutation that increases complexity will be favored by natural selection, over mutations that do not. Complexity-increasing mutations do not necessarily accumulate over time. The longest living groups of organisms on Earth are the microscopic bacteria, which have existed continuously on our planet for 3.5 billion years or so and yet exhibit no greater complexity than their old-time ancestors. More complex organisms came about later, without the elimination of their simpler relatives. Over the eons, multitudes of complex organisms have arisen on Earth. For example, the primates appeared on Earth only 50 million years ago; our species, *Homo sapiens*, less than 200,000 years ago.

Natural selection produces combinations of genes that would otherwise be highly improbable because natural selection proceeds stepwise, as illustrated by the bacterial experiment or the hypothetical typing-monkeys example. The human eye did not appear suddenly in all its present perfection. Its formation required the appropriate integration of many steps of favorable mutations; it could not have resulted from random processes alone, nor did it come about suddenly or in just a few steps. For more than half a billion years, our ancestors had some kind of organs sensitive to light. Perception of light, and later vision, were important for these organisms' survival and reproductive success. Accordingly, natural selection favored genes and gene combinations that increased the functional efficiency of the eye. Such mutations gradually accumulated, as in the evolution of the placenta in *Poeciliopsis* fish, eventually leading to the highly complex and efficient vertebrate eye. Natural selection is not by itself a creative process because it does not create the raw materials, the randomly occurring mutations. However, it becomes a creative process, causing favorable mutations to spread over multiple generations to the whole species, and by accumulating different mutations favorable to organisms over eons of time.

The Origin of Species

Although genetic mutation and natural selection are fundamental, they are only the kernel, so to speak, of the lavish tree that is the full theory of evolution. Another critical step of the evolutionary process is one called "speciation."

Speciation is largely a gradual process.[7] Yet, the history of life shows that, over time, major transitions occur, in which one kind of organism eventually becomes a very different kind. The earliest organisms on Earth were bacteria-like cells. Bacteria are microscopic and single-celled organisms, whose hereditary material is not segregated into a nucleus (hence they are called "prokaryotes"—without karyon, which is the Greek word for nucleus). Eukaryotes (having a nucleus) have cells that are larger and more complex than prokaryotic cells. Within eukaryote cells are several "organelles," where specialized functions are carried out, such as the mitochondria, where much of the cell's energy is processed, and the nucleus, which contains the hereditary information in the form of DNA organized into chromosomes. Eukaryotes have the capacity to reproduce sexually, by the fusion of two different sex cells or gametes, and most of them do. Most eukaryotes are also single-celled and microscopic.

During the first 2 billion years of life on Earth, there were only prokaryotes. Single-celled eukaryotes arose through fusion of different prokaryotes, some of which became the organelles, with their subsidiary functions, inside eukaryotes. Eventually, about 1.5 billion years ago, eukaryotic multicellular organisms appeared, with division of functions among cells—some specializing in reproduction, others becoming leaves, trunks, and roots in plants or different organs and tissues such as muscle, nerve, and bone in animals.[8] Social organization of individuals within a species is another way of achieving functional division, which may be quite fixed, as in ants and bees, or more flexible, as in cattle herds or primate colonies.

Because of the gradualness of evolution, immediate descendants differ little, and then mostly quantitatively, from their ancestors. But gradual evolution may amount to large differences over time. Sometimes, gradual morphological evolution is associated with functional changes. The forelimbs of mammals are typically adapted for walking or running, but they are adapted for shoveling earth in moles, which live mostly underground; for climbing and grasping in arboreal monkeys and apes; for swimming in dolphins and whales; and for flying in bats. The forelimbs of reptiles became wings in their bird descendants. Feathers served first for regulating temperature but eventually were co-opted for flying and became incorporated into wings.

How complex organs, such as the human eye, may arise stepwise through organs of intermediate complexity is manifest in living mollusks (squids, clams, and snails), where a gradation can be found from the simplest imaginable eye (just an eye spot consisting of a few pigmented cells with nerve fibers attached to them, as found in limpets), through a pigment cup (slit-shell mollusks), to an optic cup with a pinhole serving the role of lens (open ocean *Nautilus*), to an eye with a primitive refractive lens protected by a layer of skin cells serving as cornea (*Murex* marine snails), to the eye of octopuses and squids, as complex as the human eye, with cornea, iris, refractive lens, retina, vitreous internal substance, optic nerve, and muscle. I return to the evolution of the eye in Chapter 8.

The Baldwin Effect and Sex Determination

Natural selection proceeds in an infinite variety of ways, and yields outcomes that might seem surprising. Take the Baldwin effect, for example. In 1896, the evolutionist James M. Baldwin formulated a hypothesis that he further developed in 1902 and would later become known as the "Baldwin effect."

The hypothesis asserts that adaptive responses of organisms to extreme environments may become genetically fixed if the conditions persist. Think of adaptation to high altitude. When a person travels from near sea level to a high mountain, the body increases the production of red blood cells, since more are needed for respiration where the air is rarefied and the concentration of oxygen is low. For travelers from low to very high altitudes—for example, tourists who visit the ruins of Machu Pichu in the Andes of Peru—this adaptation occurs gradually and it will take their bodies several days to reach a satisfactory concentration of red blood cells. For people who have moved to high altitude and live there permanently, natural selection will favor genetic mutations that increase red blood cell production. This is the case for indigenous South American populations living in the high Andes: their bodies produce more red blood cells than those of people who habitually dwell at lower altitudes.

The mechanism that accounts for the Baldwin effect became known in the mid-1900s as an instance of the theoretical construct known as the "norm of reaction," which refers to the range of possible configurations that the genetic makeup of an organism can take as it becomes exposed to different environments. It was discovered that diverse adaptations in various sorts of organisms occur first as expressions of their norm of reaction, but later become fixed by natural selection promoting genetic mutations that make the adaptation permanent, if the environmental conditions persist.

The Baldwin effect has been confirmed in all sorts of organisms. Recently it has become known that often what is involved are specific switches that control gene circuits. One example that I find particularly fascinating is the chromosomal mechanism of sex determination in animals. In humans, as in other mammals, males have two different sex chromosomes, designated X and

Y, whereas females have two identical sex chromosomes, XX. In some animals, however, the sex of an individual depends on the environment. In lizards and turtles, sex is determined by the ambient temperature at which the egg develops. In some alligator species, eggs invariably produce females when incubated at up to 86°F and males when incubated at 91°F or above. Early in development, the developing sex organs (gonads) are similar in all individuals. During a critical week within the nine-week development period, the gonads differentiate into testis that produce spermatozoa (and consequently, males develop) or into an ovary with eggs (so that females develop), depending on the temperature. At temperatures between 87°F and 90°F, intermediate proportions of males and females (not hermaphrodites or intersexes) are produced.

The process is controlled by hormones. If alligator eggs developing at a male-yielding temperature (above 91°F) are exposed to the hormone estradiol, they develop into females; similarly, inhibiting estradiol yields males from eggs developing at the female-determining temperature (below 86°F). This effect is due to a circuit switch gene, *SF-1*.[9]

In mammals and birds, the mechanism determining sex is also under the control of factors like the SF-1 protein, but the production of the SF-1 protein is genetically fixed, differently in males and females. As lineages evolved from reptiles, one leading to birds and the other to mammals, sex determination became fixed in genes, so that sex was no longer dependent on the vagaries of the environment. However, the vagaries of the evolutionary process resulted in two different outcomes, one in mammals, the other in birds. In mammals the Y chromosome is a relic of an ancient X chromosome that began to lose most of its genes millions of years ago. The much-reduced Y chromosome has retained male-determining genes, so that individuals inheriting the Y chromosome invariably develop into males. In the evolution of birds, the chromosome that became smaller retained female-determining genes (rather than male-determining genes). In birds it is the females that have an unequal pair of sex-determining chromosomes (called ZW), whereas the males have two identical chromosomes (ZZ).

The Baldwin effect has been ascertained in many other interesting cases, including caste determination in social insects (ants, termites, and honeybees) and the affinity of hemoglobin for oxygen. The Baldwin effect is often involved in the origin of evolutionary novelties. Evolutionary novelties are reorganizations of preexisting morphologies, which first arise in response to environmental challenges (complex genomes have enormous plasticity), but eventually become genetically determined if the particular environmental challenges persist and the adaptation contributes importantly to survival and reproductive success.[10]

Opportunism Versus Design

An engineer has a preconception of what he wants to design and will select suitable materials and modify the design so that it fulfills the intended function. On the contrary, natural selection has no foresight, nor does it operate according to some preconceived plan. Rather, it is a purely natural process resulting from the interacting properties of physicochemical and biological entities. Natural selection is simply a consequence of the differential survival and reproduction of living beings, as pointed out. It has some appearance of purposefulness because it is conditioned by the environment: which organisms survive and reproduce more effectively depends on which variations they happen to possess that are useful or beneficial to them in the place and at the time where they live.

However, natural selection does not anticipate the environments of the future; drastic environmental changes may introduce obstacles that are insuperable to organisms that were previously thriving. In fact, species extinction is a common outcome of the evolutionary process. The species existing today represent the balance between the origin of new species and their eventual extinction. The available inventory of living species describes nearly 2 million species, but at least 10 million are estimated to exist. We also know that more than 99 percent of all species that have ever lived on Earth have become extinct without issue. Thus, since the beginning of life on Earth 3.5 billion years ago, the number of different species that have lived on our planet is likely to be more than 1 billion.

In evolution, there is no entity or person who is selecting adaptive combinations. These combinations select themselves because the organisms possessing them reproduce more effectively than those with less adaptive variations. Natural selection does not strive to produce predetermined kinds of organisms, but only organisms that are adapted to their present environments. Which characteristics will be selected depends on which variations happen to be present at a given time in a given place. This, in turn, depends on the random process of mutation, as well as on the previous history of the organisms (that is, on the genetic makeup they have as a consequence of their previous evolution). Natural selection is an "opportunistic" process, which increases the "creativity" of the process of evolution as expressed in the multiplicity and diversity of species. The variables determining the direction in which natural selection will proceed are the environment, the preexisting constitution of the organisms, and the randomly arising mutations.

Thus, adaptation to a given habitat may occur in a variety of different ways. For example, many plants have adapted to a

desert climate. Their fundamental adaptation is to the condi-
tion of dryness, which holds the danger of desiccation. During
most of the year, and sometimes for several years in succession,
there is no rain. So, plants have adapted to the urgent necessity
of saving water in different ways. Cacti have transformed their
leaves into spines, and thus avoid the evaporation that occurs
in the leaves; photosynthesis is performed on the surface of
the stem instead of in the leaves. In addition, they have transformed their stems into barrels that store a reserve of water. A
second mode of adaptation to the desert occurs in some plants
that have no leaves during the dry season, but after it rains, they
burst into leaves and flowers and quickly produce seeds. A third
mode of adaptation is that of ephemeral plants, which germi-
nate from seeds, grow, flower, and produce seeds—all within
the space of the few weeks when rainwater is available; the rest
of the year the seeds lie quiescent in the soil.

Hawaii's Evolutionary Cauldron

The opportunistic character of natural selection is also well evi-
denced by the phenomenon known as adaptive radiation. Each
of the world's continents has its own distinctive collection of
animals and plants. In Africa are rhinoceroses, hippopotamuses,
lions, hyenas, giraffes, zebras, lemurs, monkeys with narrow
noses and nonprehensile tails, chimpanzees, and gorillas. South
America, which extends over much the same latitudes as Africa,
has none of these animals; instead, it has pumas, jaguars, tapirs,
llamas, raccoons, opossums, armadillos, and monkeys with
broad noses and large prehensile tails. Australia boasts a great
diversity of marsupial mammals, which lack placentas so that
much of early development takes place in a mother's external
pouch, rather than inside the mother's womb, in the placenta.

Marsupials include the kangaroos but also moles, anteaters, and Tasmanian wolves.

The vagaries of biogeography are not due solely to the suitability of the different environments. There is no reason to believe that South American animals are not well suited to living in Africa or those of Africa to living in South America. When rabbits were intentionally introduced in Australia, so that they could be hunted for sport, they prospered beyond what the introducers had anticipated and became an agricultural pest. Hawaii lacks native land mammals, but when feral pigs and goats were introduced in the nineteenth century for hunting sports, they multiplied to large numbers and are now endangering the native vegetation.

The vagaries of biogeography (the variable distribution of organisms throughout the Earth) clearly show the opportunism of natural selection, which depends on past history, such as what organisms may or may not have colonized a territory, and on the occurrence of mutations and other chance events that open up certain evolutionary pathways and close others. Adaptive radiation, on a scale lesser than continental, is apparent in islands distant from large land masses. Darwin was startled by the Galápagos' tortoises, giant lizards, mockingbirds, and finches, different as they were from mainland species and diverse among the islands as well.

More remote yet than the Galápagos are the Hawaiian Islands, more than 2,000 miles away from the North American mainland. Many sorts of plants and animals are lacking in Hawaii, whereas others are endemic (i.e., found nowhere else on Earth) as well as extraordinarily diverse. The following table lists groups of organisms with numerous and very diverse species native to Hawaii.

	Number of Species	Percent Endemic
Ferns	168	65
Flowering plants	1,729	94
Snails	1,064	99+
Drosophila flies	510	100
Other insects	3,750	99+
Land mammals	0	0

The oldest volcano on the large island of Hawaii is some-what less than 1 million years old; Mauna Kea and Mauna Loa are much younger, and Kilauea is still active. Kauai was formed less than 10 million years ago; the other islands are of inter-mediate age, increasingly older from southeast to northwest. The gradual formation of these volcanically formed islands has resulted in successive colonizations by plants and animals, and therefore species diversification. *Drosophila* fruitflies are favored by experimental geneticists because they can easily and inexpensively be cultured in the laboratory. The ecology, behavior, and genetics of Hawaiian fruitflies have been studied intensively. There are about 1,500 known species of *Drosophila* flies in the world; nearly one-third of them live in Hawaii and nowhere else, although the total area of the archipelago is less than one-twentieth the area of California or Germany. More-over, the morphological and behavioral diversity of Hawaiian *Drosophila* exceeds that of *Drosophila* in the rest of the world. There are more than 1,000 species of land snails in Hawaii, all of which have evolved in the archipelago. There are seventy-two bird species, all but one of which exist nowhere else.

Why has such "explosive" evolution occurred in Hawaii? The overabundance of fruitflies there contrasts with the absence of many other native insects, such as mosquitoes and cock-roaches. Because of its remote isolation, the Hawaiian Islands have rarely been reached by colonizing plants and animals.

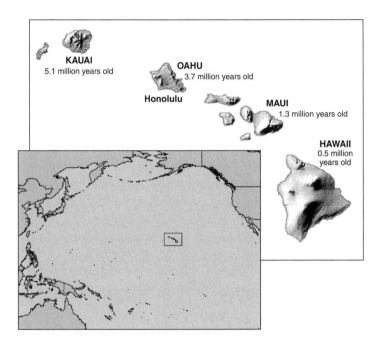

The Hawaiian Islands (with inset of the Pacific Ocean), more than 2,000 miles away from the nearest continent. These volcanic islands formed between approximately 5 million (Kauai) and 500,000 (Hawaii) years ago. (Pacific Ocean inset courtesy of NOAA Coastal Services Center.)

Some that reached the islands found suitable habitats without competitors or predators. The ancestors of Hawaiian fruitflies were passively transported to the archipelago by air currents or flotsam, before other groups of insects reached it, and there they found a multitude of opportunities for living. They rapidly evolved and diversified by exploiting the available resources. More than 500 species have derived from a single colonizing species (as is known from genetic studies); they adapted to the diversity of opportunities available in diverse niches by evolving suitable adaptations, which range broadly from one species to another. In Hawaii, some *Drosophila* species feed on decaying leaves on the forest floor; others feed on flowers; still others on fungi; and so on.

The geographic remoteness of the Hawaiian Islands is a more reasonable explanation for the explosions of diversity of a few kinds of organisms—such as fruitflies, snails, and birds—than an inordinate preference on the part of the Creator for providing the archipelago with numerous fruitflies, or a peculiar distaste for creating mosquitoes, cockroaches, and some other insects there. There are no native land mammals in Hawaii; no mammals existed there until they were introduced by humans.

Chance and Necessity

The process of natural selection can explain the design of organisms, as well as their diversity and evolution, as a consequence of their adaptation to the multifarious and ever-changing conditions of life. The fossil record shows that life has evolved in a haphazard fashion. The radiations of some groups of organisms, the numerical and territorial expansions of other groups, the replacement of some kinds of organisms by other kinds, the occasional but irregular occurrence of trends toward increased size or other sorts of change, and the ever-present extinctions are best explained by natural selection of organisms subject to the vagaries of genetic mutation, environmental challenge, and past history. The scientific account of these events does not necessitate recourse to a preordained plan, whether imprinted from the beginning or through successive interventions by an omniscient and almighty Designer. Biological evolution differs from a painting or an artifact in that it is not the outcome of preconceived design. The design of organisms is not intelligent, but imperfect and, at times, outright dysfunctional, as we shall see later.

Natural selection accounts for the "design" of organisms because adaptive variations tend to increase the probability of survival and reproduction of their carriers at the expense of maladaptive, or less adaptive, variations. The arguments of intelligent design proponents that state the incredible improbability of chance events, such as mutation, in order to account for the adaptations of organisms are irrelevant because evolution is not governed by random mutations. Rather, there is a natural process (namely, natural selection) that is not random, but oriented and able to generate order or "create." The traits that organisms acquire in their evolutionary histories are not fortuitous, but rather determined by their functional utility to the organisms, designed, as it were, to serve their life needs.

Chance is, nevertheless, an integral part of the evolutionary process. The mutations that yield the hereditary variations available to natural selection arise at random. Mutations are random or chance events because (1) they are rare exceptions to the fidelity of the process of DNA replication, and (2) because there is no way of knowing which gene will mutate in a particular cell or in a particular individual. However, the meaning of "random" that is most significant for understanding the evolutionary process is that (3) mutations are unoriented with respect to evolution; they occur independently of whether or not they are beneficial or harmful to the organisms. Some are beneficial, most are not, and only the beneficial ones become incorporated in the organisms through natural selection.

The adaptive randomness of the mutation process (as well as the vagaries of other processes that come to play in the great theater of life) is counteracted by natural selection, which preserves what is useful and eliminates what is harmful. Without hereditary mutations, evolution could not happen because there would be no variations that could be differentially conveyed from one to another generation. But without natural selection, the mutation process would yield disorganization and extinction because most mutations are disadvantageous. Mutation and selection have jointly driven the marvelous process that, starting from microscopic organisms, has yielded orchids, birds, and humans.

The time has now arrived to review the evidence for evolution, starting in Chapter 5 with the sorts of evidence that were available in Darwin's time: the fossil record; comparative anatomy; comparative embryology; and biogeography, the geographic distribution of organisms.

5

ARGUING FOR EVOLUTION

Why should the species which are supposed to have been created in the Galapagos Archipelago, and nowhere else, bear so plain a stamp of affinity to those created in America?

Charles Darwin, *The Origin of Species*, p. 398

The following assertion may bewilder some readers: Gaps of knowledge in the evolutionary history of living organisms no longer exist. This statement will come as a surprise to those who have heard again and again about "missing links," about the absence of fossil intermediates between reptiles and birds or between fish and tetrapods, and about the "Cambrian Explosion."

The evolutionary explosion that has occurred in recent years concerns knowledge, not the Cambrian: Molecular biology has made it possible to reconstruct the "universal tree of life," the continuity of succession from the original forms of life, ancestral to all living organisms, to every species now living on Earth. The main branches of the tree of life have been reconstructed on the whole and in great detail. More details about more and more branches of the universal tree of life are published in scores of scientific articles every month. The virtually unlimited evolutionary information encoded in the DNA sequence of living organisms allows evolutionists to reconstruct all evolutionary relationships leading to present-day organisms, with as much detail as wanted. Invest the necessary resources (time and laboratory expenses) and you can have the answer to any query, with as much precision as you want.

Darwin and other nineteenth-century biologists found compelling evidence for biological evolution in the comparative study of living organisms, in their geographic distribution, and in the fossil remains of extinct organisms. Since Darwin's time, the evidence from these sources has become stronger and more comprehensive, while biological disciplines that have emerged recently—genetics, biochemistry, ecology, animal behavior (ethology), neurobiology, and especially molecular biology—have supplied powerful additional evidence and detailed confirmation. Accordingly, evolutionists are no longer concerned with obtaining evidence to support the fact of evolution. Rather, evolutionary research nowadays seeks to understand further and, in more detail, how the process of evolution occurs.

In this chapter I briefly survey the sorts of knowledge available to Darwin and his contemporaries, although I update the specific evidence to the present time.[1] Notable among the immense volume of new evidence acquired since Darwin's time, are the numerous discoveries of fossil intermediates between modern humans and our ape ancestors, which are described in Chapter 6. The evidence from molecular biology is the subject of Chapter 7, where I will also describe the dramatic advances in comparative embryology, now named "evolution and development" and known as "evo-devo" for short.

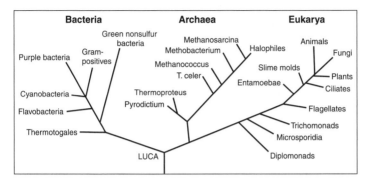

The universal tree of life, reconstructed with rRNA (ribosomal ribonucleic acid) genes. The Last Universal Common Ancestor (LUCA) is at the bottom. Branches represent different kinds of organisms. There are three major groups of organisms: bacteria, archaea, and eukaryotes. Bacteria, archaea, and most eukaryotes are microscopic. Plants, animals, and fungi are multi-cellular (macroscopic) branches of eukaryotes. (Adapted from Carl R. Woese, *Proceedings of the National Academy of Sciences* 97 (2000): 8392–8396.)

The Fossil Record

Paleontologists have recovered and studied the fossil remains of many thousands of organisms that lived in the past. These fossils show that many kinds of extinct organisms were very different in form from any now living. The fossil record also shows successions of organisms through time as well as their transition from one form to another.

When an organism dies, it is usually destroyed by bacteria and other organisms and by weathering processes. On rare occasions some body parts—particularly hard ones such as shells, teeth, and bones—are preserved by being buried in mud or protected in some other way from predators, decomposition, and weather, and they may be preserved indefinitely within the rocks in which they are embedded. (Mud and other sediments may over time become limestone and other sorts of rocks.) Methods such as radiometric dating—measuring the amounts of naturally radioactive atoms that remain in certain minerals—make it possible to estimate the time period when the rocks, and the fossils associated with them, were formed.

Radiometric dating indicates that Earth was formed about 4.5 billion years ago. The earliest fossils that resemble microorganisms such as bacteria appear in rocks more than 2.5 billion years old; some may be as old as 3.5 billion years. The oldest known animal fossils, about 700 million years old, come from the so-called Ediacara fauna, small wormlike creatures with soft bodies. Numerous fossils belonging to many animal phyla and exhibiting mineralized skeletons appear in rocks about 540 million years old, during the geological period known as the Cambrian. (A "phylum," "phyla" in plural, is a major group of organisms, such as the mollusks or the chordates.) These organisms are different from those living now and from those living at intervening times. Some are so radically different that paleontologists have had to create new phyla in order to classify them. The first vertebrates (phylum Chordata or "chordates"), animals with backbones, appeared about 400 million years ago; the first mammals, less than 200 million years ago. The history of life recorded by fossils presents compelling evidence of evolution.

The fossil record is incomplete. Of the small proportion of organisms preserved as fossils, only a tiny fraction have been recovered and studied by paleontologists; nevertheless, in some cases the succession of forms over time has been reconstructed in considerable detail. One example is the evolution of the horse, which can be traced to an animal the size of a dog having several toes on each foot and teeth appropriate for browsing

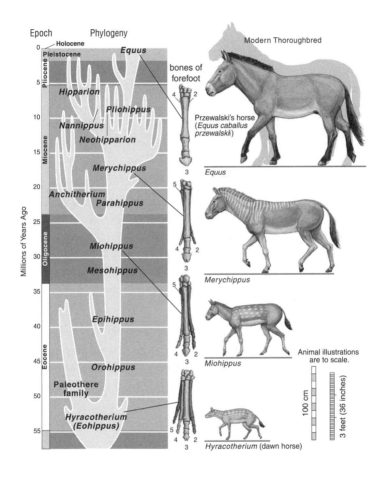

Evolution of the horse. The earliest ancestor shown is *Hyracotherium*, which lived 50 million years ago and was small, about the size of a dog. Successive species became larger, had different dentition and fewer toes, as they adapted to different diets and ways of life. Branches on the left side of the figure represent species that lived at different times, most of which became extinct. The width of the branches corresponds to the abundance of the species. (Courtesy of Encyclopaedia Britannica, Inc.)

(eating tender shoots, twigs, and leaves of trees and shrubs); this animal, called the dawn horse (scientific name *Hyracotherium*), lived more than 50 million years ago. The most recent form, the modern horse (*Equus*), is much larger, is one-toed, and has teeth appropriate for grazing (eating growing herbage). Transitional forms, as are other kinds of extinct horses that evolved in different directions and left no living descendants, are well preserved as fossils.

Using fossils, paleontologists have reconstructed examples of radical evolutionary transitions in form and function. For example, the lower jaw of reptiles consists of several bones, but that of mammals has only one. The other bones in the reptilian jaw evolved into bones now found in the mammalian ear. At first, such a transition would seem unlikely—it is hard to imagine what function such bones could have had during their intermediate stages. Yet paleontologists have discovered two transitional forms of mammal-like reptiles, called therapsids, that had a double jaw joint (i.e., two hinge points side by side)—one joint consisting of the bones that persist in the mammalian jaw and the other composed of the quadrate and articular bones, which eventually became the hammer and anvil of the mammalian ear.

Let us now examine some extinct organisms that are intermediate between different living organisms.

Archaeopteryx and *Tiktaalik*

Many fossils intermediate between diverse organisms have been discovered over the years. Two examples that have received recent attention in the media are *Archaeopteryx*, an animal intermediate between reptiles and birds, and *Tiktaalik*, intermediate between fishes and tetrapods (animals with four limbs).

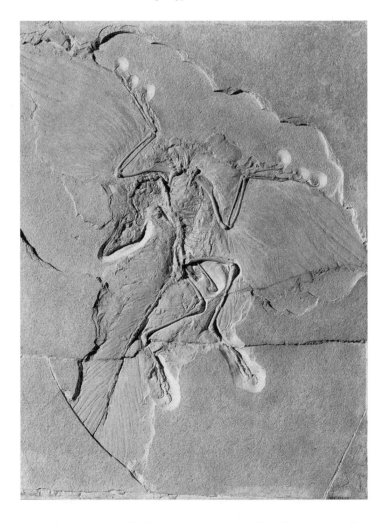

An *Archaeopteryx* fossil showing traits intermediate between reptiles (dinosaurs) and birds. (Courtesy of Museum für Naturkunde, Berlin.)

The first *Archaeopteryx* was discovered in Bavaria in 1861, two years after the publication of Darwin's *The Origin,* and received much attention because it shed light on the origin of birds and bolstered Darwin's postulate of the existence of missing links. Other *Archaeopteryx* specimens have been discovered in the past hundred years. The most recent, the tenth specimen so far recovered, was described in December 2005.[2] The best preserved *Archaeopteryx* yet, it is now housed in a small, privately owned museum in Thermopolis, Wyoming. The tetrapod-like fish *Tiktaalik* is also a very recent discovery, published only on April 6, 2006.[3]

Archaeopteryx lived during the Late Jurassic period, about 60 million years ago, and exhibited a mixture of both avian and reptilian traits. All known specimens are small, about the size of a crow, and share many anatomical characteristics with some of the smaller bipedal dinosaurs. Its skeleton is reptile-like, but *Archaeopteryx* had feathers, clearly shown in the fossils, with a skull and a beak, like those of a bird.[4]

Paleontologists have known for more than a century that tetrapods (amphibians, reptiles, birds, and mammals) evolved from a particular group of fishes called lobe-finned. Until recently, *Panderichthys* was the closest known fossil fish to the tetrapods. *Panderichthys* was somewhat crocodile shaped and had a pectoral fin skeleton and shoulder girdle intermediate in shape between those of typical lobe-finned fishes and those of tetrapods, which allowed it to "walk" in shallow waters, but probably not on land. In most features, however, *Panderichthys* was more like a fish than a tetrapod.[5] *Panderichthys* is known from Latvia, where it lived some 385 million years ago (the mid-Devonian period).

Until very recently, the earliest tetrapod fossils that are more nearly fishlike were also from the Devonian, about 376 million years old. They have been found in Scotland and Latvia.

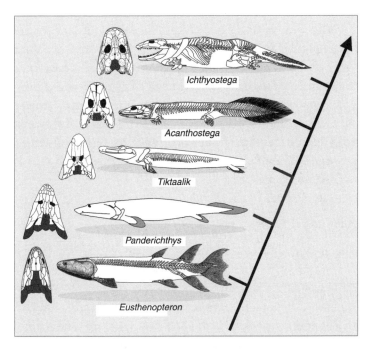

Tiktaalik and other fossil intermediates between fish and tetrapods. Other intermediate fossils, from closer to the fish to closer to the amphibians, are *Eusthenopteron, Panderichthys, Acanthastega,* and *Ichthyostega,* which lived between 385 and 359 million years ago. (Adapted from Eric Ahlberg and Jennifer A. Clark, *Nature* 440 (2006): 747. Reprinted with permission from Macmillan Publishers Ltd.)

Ichthyostega and *Acanthostega* from Greenland, which lived more recently, about 365 million years ago, are unambiguous tetrapods, with limbs that bear digits, although they retain from their fish ancestors such characteristics as true fish tails with fin rays. Thus, the time gap between the most tetrapod-like fish and the most fishlike tetrapods was nearly 10 million years, between 385 and 376 million years ago.[6]

The recently discovered *Tiktaalik* goes a long way toward breeching this gap; it is the most nearly intermediate between fishes and tetrapods yet known. Several specimens have been found in Late Devonian river sediments, dated about 380 million years ago, on Ellesmere Island in Nunavut, Artic Canada. *Tiktaalik* is Inuit for "big freshwater fish." *Tiktaalik* displays an array of features that are just about as precisely intermediate between fish and tetrapods as one could imagine and exactly fits the time gap as well.[7]

Anatomical Similarities

The skeletons of turtles, horses, humans, birds, whales, and bats are strikingly similar, in spite of the different ways of life of these animals and the diversity of their environments. The correspondence, bone by bone, can easily be seen in the limbs as well as in other parts of the body. From a purely practical point of view, it seems incomprehensible that a turtle and a whale should swim, a horse run, a person write, and a bird or bat fly with forelimb structures built of the same bones. An engineer could design better limbs for each purpose. But if we accept that all of these animals inherited their skeletal structures from a common ancestor and became modified only as they adapted to different ways of life, then the similarity of their structures makes sense.

Scientists whose field of study is comparative anatomy investigate the homologies, or inherited similarities, in the bone structure and other parts of the body between various organisms. The correspondence of structures is typically very close among some organisms—the different varieties of songbirds, for instance—but becomes less so as the organisms being compared are less closely related in their evolutionary history.

The similarities are fewer between mammals and birds than they are among mammalian species, and they are still less between mammals and fishes. Similarities in structure, therefore, not only manifest evolution but also help to reconstruct the phylogeny, or evolutionary history, of organisms.

Comparative anatomy also reveals why most organismic structures are not perfect. Like the forelimbs of turtles, horses, humans, birds, and bats, an organism's body parts are less than perfectly adapted because they are modified from an inherited structure rather than designed from completely raw materials for a specific purpose. The anatomy of animals shows that it has been designed to fit their lifestyles, but it is "imperfect" design, accomplished by natural selection, rather than "intelligent" design, as it would be if designed by an engineer. The imperfection of structures is evidence for evolution and contrary to the arguments for intelligent design.

Embryonic Development and Vestiges

Darwin and his followers found support for evolution in the study of embryology, the science that investigates the development of organisms from fertilized egg to time of birth or hatching. Vertebrates, from fishes through lizards to humans, develop in ways that are remarkably similar during early stages, but they become more and more differentiated as the embryos approach maturity. The similarities persist longer between organisms that are more closely related (e.g., humans and monkeys) than between those less closely related (such as humans and sharks).

Common developmental patterns reflect evolutionary kinship. Lizards and humans share a developmental pattern inherited from their remote common ancestor; the inherited pattern

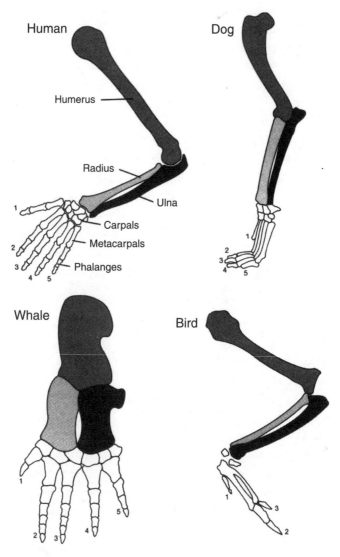

Forelimb skeleton of four vertebrates, showing similar and similarly arranged bones, although used for different functions, in human, dog, whale, and bird.

of each was modified only as the separate descendant lineages evolved in different directions. The common embryonic stages of the two creatures reflect the constraints imposed by this common inheritance, which prevents changes that have not been necessitated by their diverging environments and ways of life.

The embryos of humans and other nonaquatic vertebrates exhibit gill slits even though they never breathe through gills. These slits are found in the embryos of all vertebrates because they share as common ancestors the fish in which these structures first evolved. Human embryos also exhibit, by the fourth week of development, a well-defined tail, which reaches maximum length at six weeks. Similar embryonic tails are found in other mammals, such as dogs, horses, and monkeys; in humans, however, the tail eventually shortens, persisting only as a rudiment in the adult coccyx. Embryonic rudiments are inconsistent with claims of intelligent design: Why would some structure be designed to form during early development if it will disappear before birth? Evolution makes sense of embryonic rudiments.

A close evolutionary relationship between organisms that appear drastically different as adults can sometimes be recognized by their embryonic homologies. Barnacles, for example, are sedentary crustaceans with little apparent likeness to free-swimming crustaceans such as lobsters, shrimps, or copepods. Yet barnacles pass through a free-swimming larval stage, the nauplius, which is unmistakably similar to that of other crustacean larvae.

Embryonic rudiments that never fully develop, such as the gill slits in humans, are common in all sorts of animals. Some, however, persist as adult vestiges, reflecting evolutionary ancestry. A familiar rudimentary organ in humans is the vermiform appendix. This wormlike structure attaches to a short section of intestine called the cecum, which is located at the point

where the large and small intestines join. The human vermiform appendix is a functionless vestige of a fully developed
organ present in other mammals, such as the rabbit and other
herbivores, where a large cecum and appendix store vegeta-
ble cellulose to enable its digestion with the help of bacteria.
Vestiges are instances of imperfections—like the imperfections
seen in anatomical structures—that argue against creation by
design but are fully understandable as a result of evolution
by natural selection.

The discovery of genes, as the carriers of biological
heredity, and later of DNA as the chemical that encodes genetic
information, raised the challenge of "ontogenetic decoding,"
or the "egg-to-adult transformation." I mean by these phrases
the problem of how the linear information contained in the
sequence of letters on the DNA becomes transformed into a
three-dimensional organism that exists and changes through
time. Scientific knowledge about this problem has increased
enormously, particularly in the past three decades, made possible by the rapidly advancing discipline of molecular biology. As
noted earlier, I review this knowledge in Chapter 7, dedicated
to molecular biology.

Biogeography

Darwin saw a confirmation of evolution in the geographic
distribution of plants and animals, and later knowledge has
reinforced his observations. For example, as pointed out in
Chapter 4, there are about 1,500 known species of *Drosophila*
fruitflies in the world; nearly one-third of them live in Hawaii
and nowhere else. Also in Hawaii are more than 1,000 species
of snails and other land mollusks that exist nowhere else. This
unusual diversity is easily explained by evolution. The islands

of Hawaii are extremely isolated and have had few colonizers—that is, animals and plants that arrived there from elsewhere and established populations. Those species that did colonize the islands found many unoccupied ecological niches, local environments suited to sustaining them and lacking predators that would prevent them from multiplying. In response, the colonizing species rapidly diversified; this process of diversifying in order to fill ecological niches is called adaptive radiation.

In Chapter 4, we reviewed the remarkable diversification of life in different parts of the world that reveals the opportunism of natural selection. Even though climate and other features of the environment may be similar at similar latitudes, the flora and fauna are diverse on different continents and on different islands. This diversity occurs because natural selection depends on the opportunism of genetic mutations, which are random events. Moreover, evolution depends on previous changes, so that diversification from one continent to another, or between continents and islands, or between islands is cumulative over time. Evolutionary change occurs in response to the environment but it is conditioned by history: Mammals do not evolve into fishes, nor insects into mollusks.

Darwin's observations of the flora and fauna of South America so different from those of the Old World convinced him of the reality of evolution. The evidence from biogeography is also apparent on a scale much smaller than continental: Darwin observed that different Galápagos islands had different kinds of tortoises and different species of finches, which in turn were different from those found in continental South America. The particular case of evolution in Hawaii is a good example of how biogeography evinces evolution. Biogeography, the fanciful distribution of organisms throughout the world, can be reasonably interpreted as an outcome of evolution, rather than of the capriciousness of the Creator. As Darwin, elated by the small-scale diversity of an "entangled bank" teeming with all sorts of organisms, concluded in *The Origin*, "There is grandeur in this view of life . . . forms most beautiful and most wonderful have been, and are being evolved."

For skeptical contemporaries of Darwin, the most significant "missing link" was the absence of any known transitional form between apes and humans. No fossils were known in Darwin's time that likely would have been our ancestors after the human lineage separated from the lineage of the living apes. However, not one but many creatures intermediate between living apes and humans have since been found as fossils. Research has uncovered more and more evidence from human evolutionary history with the discovery of many fossil hominids—that is, primates belonging to the human lineage. The DNA sequence of the chimpanzee genome has been published recently. It differs little from the human genome.

In Chapter 6, I consider the fossil evidence for human evolution. I raise the question, how do relatively few gene differences account for the anatomical and behavioral differences between chimps and humans? I also raise another question, namely, how do the physicochemical signals transmitted by neurons become psychological events: feelings, thoughts, and the sense of self? We humans are the only creatures that have self-awareness, a perception of our own existence as individuals who live for a time and eventually will die.

6

Human Evolution

Man is but a reed, the weakest in nature; but he is a thinking reed.

Blaise Pascal, *Pensées*, p. 347

The missing link is no longer missing. Not one, but hundreds of fossil remains belonging to hundreds of individual hominids have been discovered since Darwin's time and continue to be discovered at an accelerated rate. The fossils that belong to the human lineage after its separation from the ape lineages are called hominids.

The oldest known fossil hominids are 6 to 7 million years old, come from Africa, and are known as *Sahelanthropus* and *Orrorin* (or *Praeanthropus*). These ancestors were predominantly bipedal when on the ground and had very small brains. *Ardipithecus* lived about 4.4 million years ago, also in Africa. Numerous fossil remains from diverse African origins are known of *Australopithecus*, a hominid that appeared between 3 and 4 million years ago. *Australopithecus* had an upright human stance but a cranial capacity of less than 500 cubic centimeters, comparable to that of a gorilla or chimpanzee and about one-third that of modern humans (500 cc are equivalent to 500 grams; 1 pound is 454 grams). The skull of *Australopithecus* displayed a mixture of ape and human characteristics—a low forehead and a long, apelike face but with teeth proportioned like those of humans. Other early hominids partly contemporaneous with *Australopithecus* include *Kenyanthropus* and *Paranthropus*; both had comparatively small brains, although some species of *Paranthropus* had larger bodies. *Paranthropus* represents a side branch of the hominid lineage that became extinct.

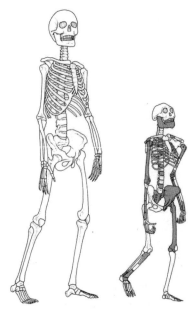

Skeleton of a modern human compared to Lucy, exemplar of *Australopithecus afarensis*, an ancestral species of modern humans that lived around 3.5 million years ago and had bipedal gait but a small brain. About 40 percent of Lucy's skeleton (shaded in the figure) was found on a single site.

Along with increased cranial capacity, other human characteristics have been found in *Homo habilis*, which lived between about 2 and 1.5 million years ago in Africa and had a cranial capacity of more than 600 cc (about 1.3 pounds), and in *Homo erectus*, which evolved in Africa sometime before 1.8 million years ago and had a cranial capacity of 800 to 1,100 cc (from nearly 2 pounds to nearly 2.5 pounds).

Homo erectus is the first intercontinental wanderer among our hominid ancestors. Shortly after its emergence in Africa, *H. erectus* spread to Europe and Asia, even as far as the Indonesian archipelago and northern China. Fossil remains of *H. erectus* have been found in Africa, Indonesia (Java), China, the Middle East, and Europe. *Homo erectus* fossils from Java have been dated at 1.81 and 1.66 million years ago, and from Georgia (in Europe, near the Asian border) between 1.6 and 1.8 million years ago.

Several species of hominids lived in Africa, Europe, and Asia between 1.8 million and 500,000 years ago. They are known as *Homo ergaster*, *Homo antecessor*, and *Homo heidelbergensis*, with brain sizes roughly that of the brain of *Homo erectus*. Some of these species were partly contemporaneous, though they lived in different regions of the Old World. These species are sometimes included under the name *Homo erectus* (in a broad sense).

The transition from *Homo erectus* to *Homo sapiens* may have started around 400,000 years ago. Some fossils of that time appear to be "archaic" forms of *H. sapiens*. Yet, *H. erectus* persisted until 250,000 years ago in China and perhaps until 100,000 years ago in Java.[1] The species *H. neanderthalensis* appeared in Europe more than 200,000 years ago and persisted until 30,000 years ago. The Neandertals have been thought to be ancestral to anatomically modern humans, but now we know that modern humans appeared more than 100,000 years ago,

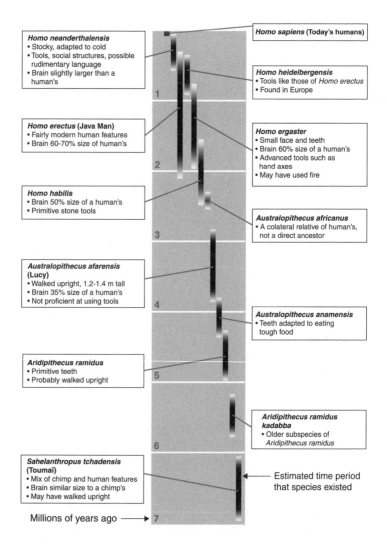

Homo neanderthalensis
• Stocky, adapted to cold
• Tools, social structures, possible rudimentary language
• Brain slightly larger than a human's

Homo sapiens (Today's humans)

Homo heidelbergensis
• Tools like those of *Homo erectus*
• Found in Europe

Homo erectus (Java Man)
• Fairly modern human features
• Brain 60-70% size of human's

Homo ergaster
• Small face and teeth
• Brain 60% size of a human's
• Advanced tools such as hand axes
• May have used fire

Homo habilis
• Brain 50% size of a human's
• Primitive stone tools

Australopithecus africanus
• A colateral relative of human's, not a direct ancestor

Australopithecus afarensis (Lucy)
• Walked upright, 1.2-1.4 m tall
• Brain 35% size of a human's
• Not proficient at using tools

Australopithecus anamensis
• Teeth adapted to eating tough food

Aridipithecus ramidus
• Primitive teeth
• Probably walked upright

Aridipithecus ramidus kadabba
• Older subspecies of *Aridipithecus ramidus*

Sahelanthropus tchadensis (Toumaï)
• Mix of chimp and human features
• Brain similar size to a chimp's
• May have walked upright

Estimated time period that species existed

Millions of years ago ⟶

Hominid species, showing several intermediate species between *Sahelanthropus tchadensis* at the bottom, which lived between 7 and 6 million years ago and modern humans at the top. (Adapted from Rex Dalton, *Nature* 440 (2006): 1101.)

much before the disappearance of Neandertal fossils. It is puzzling that, in caves in the Middle East, fossils of anatomically modern humans precede as well as follow Neandertal fossils. Some modern humans from these caves are dated at 120,000 to 100,000 years ago, whereas Neandertals are dated at 60,000 and 70,000 years, followed by modern humans dated at 40,000 years. It is unclear whether Neandertals and modern humans repeatedly replaced one another by migration from other regions, or whether they coexisted, or indeed whether interbreeding may have occurred (although comparisons of DNA from Neandertal fossils with living humans indicate that no, or very little, interbreeding occurred between Neandertals and their contemporary, anatomically modern humans).

Ancestors and Co-Lateral Relatives

Lucy is the whimsical name given to the fossil remains of a hominid ancestor classified as *Australopithecus afarensis*, a species of bipedal hominids, small brained and some 3.5 feet tall. Lucy is duly famous because about 40 percent of the whole skeleton of this young woman was found on a single site when it was discovered 30 years ago. Experts generally agree that *A. afarensis*, who lived between 3 and 3.6 million years ago, is in the line of descent to modern humans.

Australopithecus africanus, which lived more recently than *A. afarensis* and is the first *Australopithecus* species ever discovered, also was short and small-brained. However, *A. africanus* is not our ancestor, but is rather a co-lateral relative, the likely ancestor of *Australopithecus (Paranthropus) robustus* and other co-lateral hominids, who lived for 2 million years or more after their divergence from our ancestral lineage, and thus long coexisted in Africa with some of our ancestors (*A. afarensis, H. habilis,* and

H. erectus). Some of these co-lateral relatives became somewhat taller and more robust, but their brains remained small, about 500–600 cc (less than 1.5 pounds) at the most.

The discovery of hominid fossils has increased at an accelerated rate over the past two decades. In 1994, *Ardipithecus ramidus* from Ethiopia, a more primitive hominid than *Australopithecus afarensis,* was discovered, soon followed by *Australopithecus anamensis* from Kenya (dated ~3.9 to ~4.2 million years ago), as well as more specimens of *Ardipithecus* (~5.5 to 5.8 million years old) and the already mentioned *Sahelanthropus* (~6 to 7 million years old from Chad) and *Orrorin* (~5.7 to 6.0 million years old from Kenya). The position of these fossil hominids, whether in the direct ancestry of *Homo* or as co-lateral relatives remains largely a subject of debate. It has been commonly assumed, however, that *Australopithecus anamensis,* dated ~3.9 to 4.2 million years ago, is the ancestral species to *Australopithecus afarensis,* whose earliest definitive specimen is ~3.6 million years old.

The analysis and publication, on April 13, 2006, of thirty additional hominid specimens, representing a minimum of eight individuals, of *Australopithecus anamensis* from the Afar region of Ethiopia, dated to ~4.12 million years ago, confirm this interpretation.[2] The new fossils suggest, moreover, that *Ardipithecus* was the most likely ancestor of *Australopithecus anamensis* and all later australopithecines. The fossils suggest that a relatively rapid evolution from *Ardipithecus* to *Australopithecus* occurred in this region of Ethiopia.

Origin of Modern Humans

Some anthropologists have argued that the transition from *H. erectus* to archaic *H. sapiens,* and later to anatomically

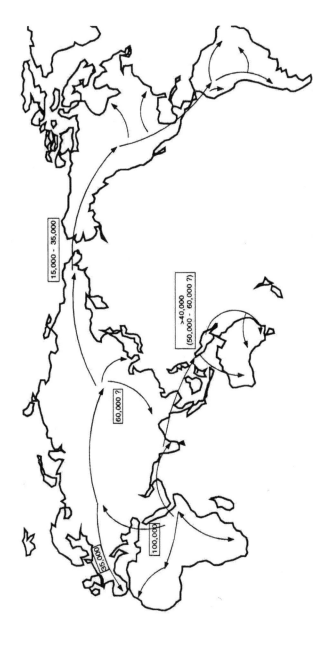

Homo sapiens colonization of the world continents, starting from its origin in tropical Africa. (From L. L. Cavalli-Sforza, P. Menozzi, and A. Piazzo, *The History and Geography of Human Genes* (Princeton, NJ: Princeton University Press, 1994), 156.)

modern humans, occurred concurrently in various parts of the Old World (Africa, Asia, and perhaps Europe). Most scientists argue instead that modern humans first arose in Africa somewhat earlier than 100,000 years ago and spread from there throughout the world, replacing the preexisting populations of *H. erectus* and related hominid species, including, later, *H. neanderthalensis.* Some proponents of this African replacement model argue further that the transition from archaic to modern *H. sapiens* was associated with a reduction of the human population to a relatively small number of individuals, and that this small number of individuals are the ancestors of all modern humans.

Recently, analyses of DNA from living humans has confirmed the African origin of modern *H. sapiens,* which is dated by these analyses at about 156,000 years ago in tropical Africa.[3] Shortly thereafter, modern humans spread through Africa and throughout the world. Southeast Asia and the region that is now China was colonized by 60,000 years ago. Shortly thereafter, modern humans reached Australasia. Europe was colonized more recently, only about 35,000 years ago, and America even more recently, perhaps only 15,000 years ago. Ethnic differentiation between modern human populations is therefore evolutionarily recent, a result of divergent evolution between geographically separated populations during the past 50,000 to 100,000 years.

Ethnicity and Race

One hundred thousand years encompass about 5,000 hominid generations, which is not a long time on the evolutionary scale. Thus, if the dispersal of modern humans from Africa to the rest of the world began 100,000 years ago, we would expect that the genetic differentiation among human populations should not be very large, even if we exclude intermingling between populations, which is occurring at an increasing rate in modern times.

Scientists have discovered that the genetic diversity among human populations of different parts of the world is only about 15 percent higher than among people from the same village.[4] This might at first seem surprising because we are aware of the conspicuously different appearance of humans from different regions of the world (the human races or ethnic groups), but it is less unexpected when we take into account the fact that the divergence of human populations is of recent origin.

The pie diagram below shows that, of the total genetic variation of all of humankind, 85 percent is present among

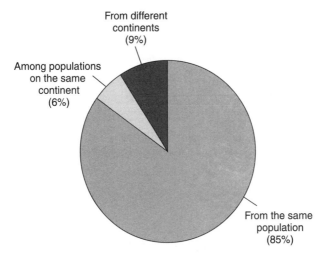

Diagram of genetic variation in human populations. Most (85 percent) of human genetic variation can be found within a single village. Populations from other villages of the same continent contribute an additional 6 percent, and those from different continents an additional 9 percent of mankind's total genetic variation.

individuals of the same population, say, of the same village or town. (This is without taking into account interbreeding with migrants from other populations, which does augment the percentage above 85.) Approximately 6 percent additional variation is found among people from different localities on the same continent, and an additional 9 percent of the variation is found among individuals from different continents.

As I have pointed out, these results may be expected because of the evolutionarily recent dispersal of human populations, but they seem to contradict common experience. We know that tropical Africans are quite different from Scandinavians and both are very different from Japanese people. The explanation of this conundrum has two components. First, our African ancestors were already genetically quite varied by the time they began to colonize the rest of the world. This is not unexpected because such is the case with most animal species: genetically they vary a great deal. Indeed, chimpanzees are genetically more varied than humans, although the total world population of chimpanzees is much smaller than the 6.5 billion humans. Second, the stereotype traits, such as skin color, hair color and texture, and facial features that distinguish ethnic groups involve relatively few genes. Some of these genes have evolved as adaptations in response to different climates. Consider, for example, one of the most conspicuous differences among ethnic groups: skin pigmentation.

Melanomas are severe cancers caused by sustained exposure to the sun's ultraviolet radiation. Thus, peoples living for generations at low latitudes have genes that produce greater amounts of eumelanins (brown and black melanin) that filter out most UV radiation and thus protect the skin from damage. On the other hand, some UV radiation is necessary for the synthesis of vitamin D in the deeper layers of the skin. Thus, the amount of eumelanin that is adaptive in the tropics is less than optimal at high latitudes, where UV radiation is much lower. At high latitudes, natural selection has favored genes that result in pale skin, so that UV reaches the layers of the dermis where vitamin D is synthesized. Examples like this one have helped demolish the myth of great genetic differentiation between "races." It is just that: a myth without scientific support.

The Ape-to-Human Transformation

Human biology in the twenty-first century faces two great research challenges: the ape-to-human and the brain-to-mind transformations. By the ape-to-human transformation, I refer to the mystery of how a particular ape lineage became a hominid lineage, from which emerged, after only a few million years, humans able to think and love, who have developed complex societies and who uphold ethical, aesthetic, and religious values. By the brain-mind transformation, I refer to the interdependent questions of (1) how the physicochemical signals that reach our sense organs become transformed into perceptions, feelings, ideas, critical arguments, aesthetic emotions, and ethical values; and (2) how, out of this diversity of experiences, there emerges a unitary reality, the mind or self. Free will and language, social and political institutions, technology and art are all epiphenomena of the human mind.

The ape-to-human and brain-to-mind transformations are major concerns for many people of faith. Are scientists claiming that humans are just another kind of ape, not any more different from chimpanzees than gorillas and other apes are? Does this imply that the religious view of humans, as special creatures of God, is without foundation? The answer to these questions is that in some biological respects we are very similar to apes, but in other *biological* respects we are very different, and these differences provide a valid foundation for a religious view of humans as special creatures of God.

The ape-to-human and the brain-to-mind transformations are intimately related questions. Scientists are far from having fully satisfactory answers to them. First, we explore how humans and apes are similar and how they differ.

Biological heredity is based on the transmission of genetic information from parents to offspring, in humans very much the same as in other animals. The DNA of humans is packaged in two sets of 23 chromosomes, one set inherited from each parent. The total number of DNA letters (the four nucleotides represented by A, C, G, T; see Chapter 7) in each set of chromosomes is about three billion, as pointed out earlier. The Human Genome Project has deciphered the sequence of the three billion letters in the human genome (i.e., in one set of chromosomes; the human genome sequence varies between genomes by about one letter in a thousand).

The two genomes (chromosome sets) of each individual are different from each other, and from the genomes of any other human being (with the trivial exception of identical twins, who share the same two sets of genes because identical twins develop from one single fertilized human egg). I estimate that the King James Bible contains about 3 million letters, punctuation marks, and spaces. Writing down the DNA sequence of one human genome demands 1,000 volumes of the size of the Bible. The human genome sequence is, of course, not printed in books, but is stored in electronic form in computers from which fragments of information can be retrieved by investigators. But if a printout is wanted, 1,000 volumes will be needed just for one human genome. Printing the complete genomic information for just one individual would demand 2,000 volumes, 1,000 for each of the two chromosome sets. Surely, again, there are more economic ways of presenting the information in the second set than listing the complete letter sequence, for example, by indicating the position of each variant letter in the second set relative to the first set. The number of variant letters between one individual's two sets is about 3 million, or 1 in 1,000.

The Human Genome Project of the United States was initiated in 1989, funded through two agencies, the National Institutes of Health and the Department of Energy. (A private enterprise, Celera Genomics, started in the United States somewhat later, but joined the government-sponsored project in achieving, largely independently, similar results.) The goal was the complete sequence of one human genome in 15 years at an approximate cost of $3 billion, coincidentally about one dollar per DNA letter. A draft of the genome sequence was completed ahead of schedule in 2001. In 2003, the Human Genome Project was finished.

Knowing the human DNA sequence is a first step, but not more than one step, toward understanding the genetic makeup of a human being. Think of the 1,000 Bible-sized volumes. We now know the orderly sequence of the 3 billion letters, but this sequence does not provide an understanding of human beings any more than we would understand the contents of 1,000 Bible-sized volumes written in an extraterrestrial language, of which we only know the alphabet, just because we would have come to decipher their letter sequence.

Human beings are not gene machines. The expression of genes in mammals takes place in interaction with the environment, in patterns that are complex and all but impossible to predict in the details—and it is in the details that the self resides. In humans, the "environment" takes a new dimension, and becomes the dominant one. Humans manipulate the natural environment so that it fits the needs of their biological makeup, for example, by making clothing and houses to live in cold climates. Moreover, the products of human technology, art, science, political institutions, and the like are dominant features of human environments.

There are two conspicuous features of human anatomy: erect posture and large brain. In mammals, brain size is generally proportional to body size. Relative to body mass, humans have the largest (and most complex) brain among all mammals. The chimpanzee's brain weighs less than a pound; a gorilla's slightly more. Our hominid ancestors had, since at least 5 million years ago, a bipedal gait, but their brain was small, little more than a pound in weight, until about 2 million years ago. Brain size started to increase notably with our *Homo habilis* ancestors, who had a brain of about a pound and a half. They became toolmakers (hence the name *habilis*) and lasted for a few hundred thousand years, starting somewhat before 2 million years ago. Adult *Homo erectus*, their descendants, had brains reaching up to somewhat more than 2 pounds in weight. An adult of our species, *Homo sapiens*, has a brain of about 3 pounds in weight, three times as large as that of the early hominids.

Our brain is not only much larger than that of chimpanzees or gorillas, but also much more complex. Our cerebral cortex, where the higher cognitive functions are processed, takes up a disproportionally much greater part of our brain than that of the apes.

A draft of the DNA sequence of the chimpanzee genome was published on September 1, 2005. In the genome regions shared by humans and chimpanzees, the two species are 99 percent identical. The differences appear to be very small or quite large, depending on how one chooses to look at them: One percent of the total seems very little, but it amounts to a difference of 30 million DNA letters out of the 3 billion in each genome. Twenty-nine percent of the enzymes and other proteins encoded by the genes are identical in both species. Out of the one hundred to several hundred amino acids that make up each protein, the 71 percent of nonidentical proteins differ between humans and chimps by only two amino acids, on the average. If one takes into account DNA segments found in one species but not the other, the two genomes are about 96 percent identical, rather than nearly 99 percent identical as in the case of DNA sequences shared by both species. That is, a large amount of genetic material—about 3 percent, or some 90 million DNA letters—has been inserted or deleted since humans and chimps initiated their separate evolutionary ways, 7 to 8 million years ago. Most of this DNA does not contain genes coding for proteins.

Comparison of the two genomes provides insights into the rate of evolution of particular genes in the two species. One significant finding is that genes active in the brain have changed more in the human lineage than in the chimp lineage. Also significant is that the fastest evolving human genes are those coding for "transcription factors," that is, "switch" proteins, which control the expression of other genes. They determine when other genes are turned on and off. On the whole, 585 genes, including genes involved in resistance to malaria and tuberculosis, have been identified as evolving faster in humans than in chimps. (Note that malaria is a severe disease for humans but not for chimps.)

Genes located on the Y chromosome, found only in the male, have been much better protected by natural selection in the human than in the chimpanzee lineage, in which several genes have incorporated disabling mutations that make the genes nonfunctional. Also, there are several regions of the human genome that seem to contain beneficial genes that have rapidly evolved within the past 250,000 years. One region contains the *FOXP2* gene, involved in the evolution of speech.

All this knowledge (and much more of the same kind that will be forthcoming) is of great interest, but what we so far know advances but very little our understanding of what genetic changes make us distinctively human. However, we know some basic features that account for human distinctness and therefore can serve as foundations for a religious view of humankind: the large brain and the accelerated rate of evolution of genes, such as those involved in human speech.

Extended comparisons of the human and chimp genomes and experimental exploration of the functions associated with significant genes will surely advance considerably our understanding, over the next decade or two, of what it is that makes us distinctively human. Surely also, full biological understanding will only come if we also solve the second conundrum, the brain-to-mind transformation, which I identified earlier. The distinctive features that make us human begin early in development, well before birth, as the linear information encoded in the genome gradually becomes expressed into a four-dimensional individual, an individual who changes in configuration as time goes by. In an important sense, the most distinctive human features are those expressed in the brain, those that account for the human mind and for human identity.

Some Christian believers will say that the fundamental difference between humans and apes is that we have a soul, created by God, which the apes do not have. This is a religious or theological answer that will be satisfying for many believers, but it is not *scientifically* satisfactory. What I mean is that, soul or no soul, scientists still want to learn how the anatomical and behavioral differences between humans and apes come to emerge from genetic differences between them. Surely, believers in the soul would not, I hope, believe that there are no biological correlates that account for the ape-to-human differences. That is what scientists seek to understand: what are the genetic and other features that distinguish our species from apes and other animals. Consider, by analogy, a human individual. People of faith may believe that it is the soul infused by God that accounts for what each person is. But surely, this does not deny that each individual develops from a fertilized egg in the mother's womb and later by multiple cell divisions. Nor will we want to ignore the genetic and other features that distinguish one person from another.

As biological understanding of the differences between humans and apes advances, there will surely be much left for philosophical reflection, as well as plenty of issues with great theological significance. Biological knowledge does not eliminate religious belief. Rather, scientific knowledge may provide a basis for theological insights.

The Brain-to-Mind Puzzle

The brain is the most complex and most distinctive human organ. It consists of 30 billion nerve cells, or neurons, each connected to many others through two kinds of cell extensions, known as axons and dendrites. From the evolutionary point of view, the animal brain is a powerful biological adaptation; it allows an organism to obtain and process information about environmental conditions and then to adapt to them. This ability has been carried to the limit in humans, in which the extravagant hypertrophy of the brain makes possible abstract thinking, language, and technology. By these means, humankind has ushered in a new mode of adaptation far more powerful than the biological mode: adaptation by culture (see below).

The most rudimentary ability to gather and process information about the environment is found in certain single-celled microorganisms. The protozoan *Paramecium* swims, apparently at random, ingesting the bacteria it encounters, but when it meets unsuitable acidity or salinity, it checks its advance and starts in a new direction. The single-celled alga *Euglena* not only avoids unsuitable environments but seeks suitable ones by orienting itself according to the direction of light, which it perceives through a light-sensitive spot in the cell. Plants have progressed further, but not very far. Except for those with tendrils that twist around any solid object and the few carnivorous plants that react to touch, plants mostly react only to gradients of light, gravity, and moisture.

In animals the ability to secure and process environmental information is mediated by the nervous system. The simplest nervous systems are found in corals and jellyfishes; they lack coordination between different parts of their bodies, so any one part is able to react only when it is directly stimulated. Sea urchins and starfish possess a nerve ring and radial nerve cords that coordinate stimuli coming from different parts; hence, they respond with direct and unified actions of the whole body. They have no brain, however, and seem unable to learn from experience. Planarian flatworms have the most rudimentary brain known; their central brain and nervous system process and coordinate information gathered by sensory cells. These animals are capable of simple learning and hence of variable responses to repeatedly encountered stimuli. Insects and their relatives have yet more advanced brains; they obtain precise chemical, acoustic, visual, and tactile signals from the environment and process them, making possible complex behaviors, particularly in the search for food, selection of mates, and social organization.

Vertebrates—animals with backbones—are able to obtain and process much more complicated signals and to respond to the environment more variably than insects or any other invertebrates. The vertebrate brain contains an enormous number of associative neurons arranged in complex patterns. In vertebrates the ability to react to environmental information increases with increasing size of the cerebral hemispheres and of the neopallium, an organ that associates and coordinates signals from all receptors and brain centers. In mammals, the neopallium has expanded and become the cerebral cortex. Humans have a very large brain relative to their body size, and a cerebral cortex that is disproportionately large and complex even for their brain size. Abstract thinking, symbolic language, complex social organization, values, ethics, and religion are manifestations of the wondrous capacity of the human brain to gather information about the external world and to integrate that information and react flexibly to what is perceived.

Cultural Evolution

With the advanced development of the human brain, biological evolution has transcended itself, opening up a new mode of evolution: adaptation by technological manipulation of the environment. Organisms adapt to the environment by means of natural selection, by changing their genetic constitution over the generations to suit the demands of the environment. Humans (and humans alone, at least to any significant degree), have developed the capacity to adapt to hostile environments by modifying the environments according to the needs of their genes. The discovery of fire and the fabrication of clothing and shelter have allowed humans to spread from the warm tropical and subtropical regions of the Old World, to which we are biologically adapted, to almost the whole Earth; it was not necessary for wandering humans that they wait until genes providing anatomical protection against cold temperatures by means of fur or hair would evolve. Nor are we humans biding our time in expectation of wings or gills; we have conquered the air and seas with artfully designed contrivances—airplanes and ships. It is the human brain (or rather, the human mind) that has made humankind the most successful, by most meaningful standards, living species.

There are not enough bits of information in the complete DNA sequence of a human genome to specify the trillions of connections among the 30 billion neurons of the human brain. Accordingly, the genetic instructions must be organized in control circuits operating at different hierarchical levels (see Chapter 7), so that an instruction at one level is carried through many channels to lower levels in the hierarchy of control circuits.

One of the most exciting biological disciplines that has made great strides within the past two decades is neurobiology. An increased commitment of financial and human resources to that field has enabled an unprecedented rate of discovery. Much has been learned about how light, sound, temperature, resistance, and chemical impressions received in our sense organs trigger the release of chemical transmitters and electric potential differences that carry the signals through the nerves to the brain and elsewhere in the body. Much has also been learned about how neural channels for information transmission become reinforced by use or may be replaced after damage; about which neurons or groups of neurons are committed to processing information derived from a particular organ or environmental location; and about many other issues concerning neural processes. But, for all this progress, neurobiology

remains an infant discipline, at a stage of theoretical development comparable perhaps to that of genetics at the beginning
of the twentieth century. Those things that count most remain
shrouded in mystery: how physical phenomena become mental
experiences (the feelings and sensations, called "qualia" by
philosophers, that contribute the elements of consciousness),
and how out of the diversity of these experiences emerges the
mind, a reality with unitary properties, such as free will and the
awareness of self that persist through an individual's life.[5]

Believers might say, once again, that the soul accounts for
the mind. Once again, however, scientists will want to understand the biological (as well as chemical and electrical) correlates that account for mental experiences.[6] I do not believe that
the mysteries of the mind are unfathomable; rather, they are
puzzles that humans can solve with the methods of science and
illuminate with philosophical analysis and reflection. And I will
place my bets that, over the next half century or so, many of
these puzzles will be solved and that they will provide valuable
religious and theological insights. We shall then be well on our
way toward answering the injunction: "Know thyself."

In the next chapter, we complete our review of the evidence
for evolution by looking into molecular biology, a young scientific discipline that provides the strongest evidence yet that
evolution has occurred. Molecular biology, moreover, makes it
possible to reconstruct the evolutionary relationships among
living organisms with as much precision as desired.

7

MOLECULAR BIOLOGY

The aim of molecular biology is to interpret biological structures and performances in explicitly molecular terms.

P. B. and J. S. Medawar, *Aristotle to Zoos*, p. 186

The universal tree of life embraces all living organisms from
their common ancestor to the present. In the figure on page 81,
groups of organisms are represented by the branches of the tree.
There are three major sets of branches: eukaryotes, bacteria,
and archaea; the last two are prokaryotes and are microscopic
organisms. Most eukaryotic organisms are also microscopic
single cells, but the familiar animals, plants, and fungi are multicellular organisms; they are represented as three of the many
branches of the eukaryotes. All organisms are related by common descent from a single form of life, represented by the tree's
"trunk" (the straight-up line at bottom).[1]

Molecular biology, a discipline that emerged in the second
half of the twentieth century, nearly 100 years after the publication of *The Origin of Species*, has provided the strongest
evidence yet of the evolution of organisms. Molecular biology
proves evolution in two ways: first, by showing the unity of life

in the nature of DNA and the workings of organisms at the level of enzymes and other protein molecules; second, and most important for evolutionists, by making it possible to reconstruct evolutionary relationships that were previously unknown, and to confirm, refine, and time all evolutionary relationships from the universal common ancestor up to all living organisms. The precision with which these events can be reconstructed is one reason why the evidence from molecular biology is so useful to evolutionists and so compelling.

The Unity of Life

The molecular components of organisms are remarkably uniform—in the nature of the components as well as in the ways in which they are assembled and used. In all bacteria, archaea, plants, animals, and humans, the instructions that guide the development and functioning of organisms are encased in the same hereditary material, DNA, which provides the instructions for the synthesis of proteins. The thousands of enormously diverse proteins that exist in organisms are synthesized from different combinations, in sequences of variable length, of twenty amino acids, the same in all proteins and in all organisms. Yet several hundred other amino acids exist. Moreover, the genetic code, by which the information contained in the DNA of the cell nucleus is passed on to proteins, is virtually everywhere the same.[2] Similar metabolic pathways—sequences of biochemical reactions—are used by the most diverse organisms to produce energy and to make up the cell components. Many other pathways are theoretically possible, but only a limited number are used in organisms, and the pathways are the same in organisms with extremely different ways of life.

The unity of life reveals the genetic continuity and common ancestry of all organisms. There is no other rational way to account for their molecular uniformity, given that numerous alternative structures and fundamental processes are in principle equally likely. The genetic code may serve as an example. Each particular sequence of three nucleotides (called a "triplet" or "codon") in the nuclear DNA acts as a code for exactly the same particular amino acid in all organisms. For example, in any given gene of any organism, the codon GCC determines that the amino acid alanine will be incorporated in the protein specified by the gene; the codon GAC determines the incorporation of the amino acid asparagine, and so on.[3] The universal correspondence between the DNA language (codons) and the protein language (amino acids) is no more necessary than it is for any two spoken languages to use the same combination of letters for representing the same particular concept or object. If we find that certain sequences of letters—planet, tree, woman—are used with identical meanings in a number of different books, we can be sure that the languages used in the books are identical, and that they must have had a common origin.

Genetic Information

Genes and proteins are long molecules that contain information in the sequence of their components in a way similar to how the meaning of sentences in any language is conveyed by the sequence of letters and words. The sequences that make up the genes are passed on from parents to offspring and are identical from generation to generation, except for occasional changes introduced by mutations. Closely related species have very similar DNA sequences; the few differences reflect mutations that have occurred since their last common ancestor.

Species that are less closely related to one another exhibit more differences in their DNA than those more closely related because more time has elapsed since their last common ancestor. This is the rationale used for reconstructing evolutionary history using molecules, which may be illustrated with the following analogy.

Let us assume that we are comparing two books written in the same language. Both books are 200 pages long and contain the same number of chapters. Closer examination reveals that the two books are identical page for page and word for word, except that an occasional word—say, 1 in 100—is different. The two books cannot have been written independently; either one has been copied from the other, or both have been copied, directly or indirectly, from the same original book. In living beings, if each component nucleotide of DNA is represented by one letter, the complete sequence of nucleotides in the DNA of a higher organism would require several hundred books, each with hundreds of pages, with several thousand letters on each page. When the "pages" (or sequences of nucleotides) in these "books" (genomes) are examined one by one, the correspondence in the "letters" (nucleotides) gives unmistakable evidence of common origin. But we can go one step further.

As pointed out above, molecular biology offers two kinds of arguments for evolution. Using the alphabet analogy, the first argument says that languages that use the same alphabet (the same hereditary molecule, the DNA made up of the same four nucleotides, and the same 20 amino acids in their proteins) as well as the same dictionary (the same genetic code) cannot be of independent origin. The second argument concerns the degree of similarity in the sequence of nucleotides in the DNA (and thus the sequence of amino acids in the proteins); it says that books with very similar texts cannot be of independent origin. The degree of similarity between sequences of DNA (and protein) is what makes it possible to reconstruct evolutionary history. We'll return below to this second kind of evidence (see "Informational Macromolecules").

From Mendel to Dolly

The theory of biological heredity was formulated by the Augustinian monk Gregor Mendel in 1866, but it became generally known by biologists only in 1900: genetic information is contained in discrete factors, or genes, which exist in pairs, one member of the pair received from each parent. The next step toward understanding the nature of genes was completed during the first quarter of the twentieth century. It was established that genes are parts of the chromosomes and that they are arranged linearly along the chromosomes. It took another quarter century to determine the chemical composition of genes—DNA. As pointed out earlier, DNA consists of four kinds of chemical components (nucleotides) organized in long, double-helix-shaped molecules.[4] The genetic information is contained in the linear sequence of the nucleotides, very much the way semantic information of an English sentence is conveyed by the particular sequence of the twenty-six letters of the English alphabet.

The first important step toward understanding how the genetic information is decoded came in 1941 when George W. Beadle and Edward L. Tatum demonstrated that genes determine the synthesis of enzymes; enzymes are the catalysts that control all chemical reactions in living beings. Later it became known that amino acids (the components that make up enzymes and other proteins) are each encoded by a set of three consecutive nucleotides (codon). This relationship explains the linear correspondence between a particular sequence of coding nucleotides in DNA and the sequence of the amino acids that make up the encoded enzyme.

Chemical reactions in organisms must occur in an orderly manner; organisms must have ways of switching genes on and off, since different sets of genes are active in different cells. The first gene control system was discovered in 1961 by François Jacob and Jacques Monod for a gene that encodes an enzyme that digests sugar in the bacterium *Escherichia coli*. The gene is turned on and off by a system of several switches consisting of short DNA sequences adjacent to the coding part of the gene. (The coding sequence of a gene is the part that determines the sequence of amino acids in the encoded enzyme.) The switches acting on a given gene are activated or deactivated by feedback loops that involve molecules synthesized by other genes. A variety of gene control mechanisms were soon discovered, in bacteria and other microorganisms. Two elements typically are present in gene control: feedback loops and short DNA sequences that act as switches. The feedback loops ensure that the presence of a substance in the cell induces the synthesis of the enzyme required to digest it, and that an excess of the enzyme in the cell represses its own synthesis. For example, the gene encoding a sugar-digesting enzyme in *E. coli* is turned on or off by the presence or absence of the sugar lactose.

The investigation of gene control mechanisms in mammals (and other complex organisms) became possible in the mid-1970s with the development of recombinant DNA techniques. This technology made it feasible to isolate single genes (and other DNA sequences) and to multiply, or "clone," them, in billions of identical copies, in order to obtain the quantities necessary for ascertaining their nucleotide sequence. One unanticipated discovery was that most genes come in pieces: the coding sequence of a gene is divided into several fragments separated one from the next by noncoding DNA segments. In addition to the alternating succession of coding and noncoding segments, mammalian genes contain short control sequences, like those in bacteria but typically more numerous and complex, that act as control switches and signal where the coding sequence begins.

Much remains to be discovered about the control mechanisms of mammalian genes. The daunting speed at which molecular biology is advancing has enabled the discovery of some prototypes of mammalian gene control systems, but much remains to be unraveled. Moreover, understanding the control mechanisms of individual genes is but the first major step toward solving the mystery of "ontogenetic decoding," that is, how DNA determines the characteristics of fully developed multicellular organisms, such as plants and animals. The second major step is solving the puzzle of differentiation.

Egg to Adult

A human being consists of 1 trillion cells of some 300 different kinds, all derived by sequential division, each cell dividing into two, from the fertilized egg, a single cell 0.1 millimeter (about four thousandths of an inch) in diameter. The first few cell divisions yield a spherical mass of undifferentiated cells. Successive divisions see the cells starting to differentiate with the appearance of folds and ridges in the mass of cells and, later, of the variety of tissues, organs, and limbs characteristic of a human individual. The full complement of genes duplicates with each cell division, so that two complete genomes are present in every cell. Yet different sets of genes are active in different cells. This must be so in order for cells to differentiate: a nerve cell, a muscle cell, and a skin cell are vastly different in size, configuration, and function. The differential activity of genes must continue after differentiation because different cells fulfill different functions, which are controlled by different genes. Nevertheless, experiments with other animals (and some with humans) indicate that all the genes in any cell have the potential to become activated. (The sheep Dolly was conceived using the genes extracted from a cell in an adult sheep.)

The information that controls cell and organ differentiation is ultimately contained in the DNA sequence, but mostly in very short segments of it. In mammals, insects, and other complex organisms, there are control circuits that operate at higher levels than the control mechanisms that activate and deactivate individual genes. These higher-level circuits (such as the so-called *homeobox*, or *hox*, genes) act on sets of genes rather than individual ones. The details of how these sets are controlled, how many control systems there are, and how they interact, as well as many other related questions, are what need to be resolved to elucidate the egg-to-adult transformation, how a fertilized egg develops into a mature individual. The DNA sequence of some controlling elements has been ascertained, but this is a minor success that is only helped a little by plowing through the entire 3 billion nucleotide pairs that constitute the human genome. Experiments with stem cells are likely to provide important knowledge as scientists ascertain how stem cells become brain cells in one case, muscle cells in another, and so on.

The benefits that the elucidation of ontogenetic decoding will bring to mankind are enormous. This knowledge will make it possible to understand the modes of action of complex genetic diseases, including cancer, and therefore their cures. It will also bring an understanding of the process of aging, the unforgiving infirmity that kills all those who have won the battle against other maladies.

Cancers are anomalies of ontogenetic decoding: cells proliferate although the welfare of the organism demands otherwise. Individual genes ("oncogenes") have been identified that are involved in the causation of particular forms of cancer. Whether or not a cell will turn out cancerous, however, depends on the interaction of the oncogenes with other genes and with the internal and external environment of the cell. Aging is a failure of the process of ontogenetic decoding: Cells fail to carry out the functions imprinted in their genetic codescript or are no longer able to proliferate and replace dead cells.

Informational Macromolecules

DNA and proteins have been called "informational macromolecules" because they are long linear molecules made up of sequences of units—nucleotides in the case of nucleic acids, amino acids in the case of proteins—that embody evolutionary information. Comparing the sequence of the components in two macromolecules establishes how many units are different. Because evolution usually occurs by changing one unit at a time, the number of differences is an indication of the recency of common ancestry.

The degree of similarity in the sequence of nucleotides, or of amino acids, can be precisely quantified. For example, in humans and chimpanzees, the protein molecule called cytochrome *c*, which serves a vital function in respiration within cells, consists of the same 104 amino acids in exactly the same order. It differs, however, from the cytochrome *c* of rhesus monkeys by one amino acid, from that of horses by eleven additional amino acids, and from that of tuna by twenty-one additional amino acids.

The degree of similarity reflects the recency of common ancestry. Thus, the inferences from paleontology, comparative anatomy, and other disciplines that study evolutionary history can be tested in molecular studies of DNA and proteins by examining the sequences of nucleotides and amino acids. The authority of this kind of test is overwhelming: each of the thousands of genes and thousands of proteins contained in an organism provides an independent test of that organism's evolutionary history.

Molecular evolutionary studies have three notable advantages over comparative anatomy and the other classical disciplines. One is that the information is readily quantifiable. The number of units that are different is easily established when the sequence of units is known for a given macromolecule in different organisms. It is simply a matter of aligning the units (nucleotides or amino acids) between two or more species and counting the differences. The second advantage is that comparisons can be made between very different sorts of organisms. There is very little that comparative anatomy can say when, for example, organisms as diverse as yeasts, pine trees, and human beings are compared, but there are numerous DNA and protein sequences that can be compared in all three. The third advantage is multiplicity. Each organism possesses thousands of genes and proteins, every one of which reflects the same evolutionary history. If the investigation of one particular gene or protein does not satisfactorily resolve the evolutionary relationship of a set of species, additional genes and proteins can be investigated until the matter has been settled.

Moreover, the widely different rates of evolution of different sets of genes opens up the opportunity for investigating different genes in order to achieve different degrees of resolution in the tree of evolution. Evolutionists rely on slowly evolving genes for reconstructing remote evolutionary events, but increasingly faster evolving genes for reconstructing the evolutionary history of more recently diverged organisms.

Genes that encode *ribosomal RNA* molecules are among the slowest evolving genes. (Ribosomes are complex molecules that mediate the synthesis of proteins; each ribosome consists of several proteins and several RNA molecules.) They have been used to reconstruct the evolutionary relationships among groups of organisms that diverged very long ago: for example, among bacteria, archaea, and eukaryotes (the three major divisions of the living world), which diverged more than 2 billion years ago (see the figure on page 81), or among the microscopic protozoa (e.g., *Plasmodium*, which causes malaria) compared with plants and with animals, groups of organisms that diverged about 1 billion years ago. Cytochrome c, mentioned earlier, evolves slowly, but not as slowly as the ribosomal RNA genes. Thus, it is used to decipher the relationships within large groups of organisms, such as among humans, fishes, and insects. Fast-evolving molecules, such as the fibrinopeptides involved in blood clotting, are appropriate for investigating the evolution of closely related animals—the primates, for example: macaques, chimps and humans.

Lineage Evolution and Species Diversification

DNA and proteins provide information not only about the branching of lineages from common ancestors (diversification, called "cladogenesis") but also about the amount of genetic change that has occurred in any given lineage (known as "anagenesis"). It might seem at first that quantifying anagenesis for proteins and nucleic acids would be impossible because it seems to require comparison of molecules from organisms that are now extinct with molecules from living organisms. Organisms of the past are sometimes preserved as fossils, but their DNA and proteins have largely disintegrated. Nevertheless, comparisons between living species provide information about anagenesis.

Consider, for example, the protein cytochrome *c*, involved in cell respiration. The sequence of amino acids in this protein is known for many organisms, from bacteria and yeasts to insects and humans; in animals, cytochrome *c* consists of 104 amino acids. When the amino acid sequences of humans and rhesus monkeys are compared, they are found to be different at position 58: isoleucine (I) in humans, threonine (T) in rhesus monkeys, but identical at the other 103 positions (see figure below). When humans are compared with horses, twelve amino

Human   G – D – V – E – K – G – K – K – I – F – I – M – K – C – S – Q – C –
Rhesus monkey   • • • • • • • • • • • • • • • •
Horse   • • • • • • • • • • V Q • • A • •

H – Y – V – E – K – G – G – K – H – K – Y – G – P – N – L – H – G – L – F – G – R – K – T –
• •
• •

G – Q – A – P – G – Y – S – Y – T – A – A – N – K – N – K – G – I – I – W – G – E – D – T –
• • • • • • • • • • • • • • • • • T • • • •
• • • • • F T • • D • • • • • • • T • K • E •

L – M – E – Y – L – E – N – P – K – K – Y – I – P – G – T – K – M – I – F – V – G – I – K –
• •
• • • • • • • • • • • • • • • • • • A • • •

K – K – E – E – R – A – D – L – I – A – Y – L – K – K – A – Y – N – E
• • • • • • • • • • • • • • • • • •
• • T • • E • • • • • • • • • • • •

The 104 amino acids in the cytochrome *c* of human are shown on top (using standard one-letter representations for each amino acid). At one position rhesus monkeys have threonine, while humans have isoleucine. Humans differ from horse by twelve amino acids; monkey and horse differ by eleven amino acids. Dots indicate amino acids identical to those of human cytochrome *c*. (After Fitch and Margoliash, *Science* 155 (1967): 279–284).

acid differences are found, and when horses are compared with rhesus monkeys, there are eleven amino acid differences. Even without knowing anything else about the evolutionary history of mammals, we would conclude that the lineages of humans and rhesus monkeys diverged from each other much more recently than they diverged from the horse lineage. Moreover, it can be concluded that the amino acid difference between humans and rhesus monkeys must have occurred in the human lineage after its separation from the rhesus monkey lineage (see figure below).

Evolutionary trees are models or hypotheses that seek to reconstruct the evolutionary history of taxa—that is, species, genera, families, orders, and other groups of organisms. As I have

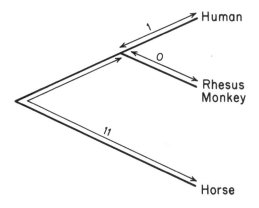

Evolutionary tree of human, rhesus monkey, and horse, based on their cytochrome *c.* The one difference between human and monkey (figure on preceding page) is due to a change in the human lineage, after it diverged from the monkey lineage. This conclusion is reached because monkey and horse (as well as other animals) have the same amino acid (T) at this position, while the human is different (amino acid I).

pointed out above, the trees embrace two kinds of information related to evolutionary change, cladogenesis and anagenesis. The figure on page 129 illustrates both. The branching relationships of the tree reflect the relative relationships of ancestry, or cladogenesis. Thus, in the figure, humans and rhesus monkeys are seen to be more closely related to each other than either one is to the horse. Stated another way, this tree shows that the last ancestor of all three species lived in a more remote past than the last ancestor of humans and monkeys.

Evolutionary trees also indicate the changes that have occurred along each lineage, or anagenesis. Thus, in the evolution of cytochrome *c* since the last common ancestor of humans and rhesus monkeys, one amino acid has changed in the lineage going to humans but none in the lineage going to rhesus monkeys. This conclusion is drawn from the observation that, at one position, monkeys and horses (as well as other animals, see below) have the same amino acid (threonine), whereas humans have a different one (isoleucine), which therefore must have changed in the human lineage after it separated from the monkey lineage. The amino acid sequences in the cytochrome *c* of twenty very diverse organisms were ascertained in 1967.[5] Counting the amino acid differences between the twenty species resulted in the evolutionary tree shown in the figure.

The reconstruction of evolutionary history accomplished with DNA and protein molecules follows the same logic used in comparative anatomy and other traditional methods: the degree of similarity reflects the recency of common ancestry. In paleontology, the time sequence of fossils is determined by the age of the rocks in which they are embedded. The inferences from comparative anatomy, paleontology, and other disciplines pertaining to evolutionary history can be tested in molecular studies of DNA and proteins by examining the sequences of

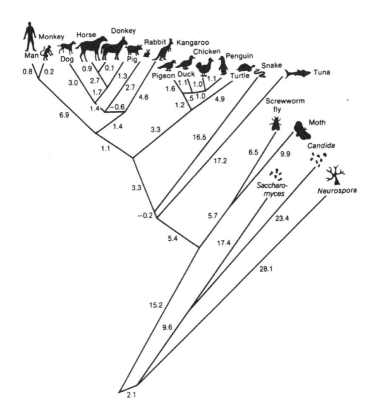

Evolutionary history of twenty species, based on the cytochrome *c* amino acid sequence. The common ancestor (at the bottom) of yeast and humans lived more than 1 billion years ago. The numbers along the branches estimate the nucleotide substitutions occurring in the span of evolution represented by the branch. Although fractional (or negative) numbers of nucleotide substitutions cannot occur, the numbers along the branches are those that best fit the data. More detailed studies would make it possible to determine the exact number of changes along each branch. (After Fitch and Margoliash, *Science* 155 (1967): 279–284.)

nucleotides and amino acids. As pointed out above, each of the thousands of genes and thousands of proteins contained in an organism provides an independent test of that organism's evolutionary history. Many thousands of tests have been done (and thousands more are published every year); not one has given evidence contrary to evolution. There is probably no other notion in any field of science that has been as extensively tested and as thoroughly corroborated as the evolutionary origin of living organisms.

The Molecular Clock

One conspicuous attribute of molecular evolution is that differences between homologous molecules can readily be quantified and expressed as, for example, proportions of nucleotides or amino acids that have changed. (Homologous molecules are those derived from a common ancestral molecule.) Rates of evolutionary change can therefore be more precisely established with respect to DNA or proteins than with respect to traits, such as the configuration or function of an organ or limb. Studies of molecular evolution rates have led to the proposition that macromolecules may serve as evolutionary clocks.

It was first observed in the 1960s that the number of amino acid differences between homologous proteins of any two given species seemed to be nearly proportional to the time of their divergence from a common ancestor. If the rate of evolution of a protein or gene were approximately the same in the evolutionary lineages leading to different species, proteins and DNA sequences would provide a molecular clock of evolution. The sequences could then be used to reconstruct not only the sequence of branching events of a phylogeny but also the time when the various events occurred.

Consider, for example, the figure on page 131. If the substitution of nucleotides in the gene coding for cytochrome *c* occurred at a constant rate through time, we could determine the time elapsed along any branch of the phylogeny simply by examining the number of nucleotide substitutions along that branch. We would need only to calibrate the clock by reference to an outside source, such as the fossil record, that would provide the actual geologic time elapsed in at least one specific lineage or since one branching point. For example, if the time of divergence between insects and vertebrates is determined to have occurred 700 million years ago, other times of divergence can be determined by proportion of the number of amino acid changes.

The molecular evolutionary clock is not expected to be a metronomic clock, like a watch or other timepieces that measure time exactly, but a stochastic (probabilistic) clock, like radioactive decay. In a stochastic clock the probability of a certain amount of change is constant (e.g., a given quantity of atoms of carbon-14 is expected, through decay, to be reduced by half in 5,730 years, its "half-life"), although some variation occurs in the actual amount of change. Over fairly long periods of time a stochastic clock can be quite accurate. The enormous potential of the molecular evolutionary clock lies in the fact that each gene or protein is a separate clock. Each clock "ticks" at a different rate—the rate of evolution characteristic of a particular gene or protein—but each of the thousands and thousands of genes or proteins provides an independent measure of the same evolutionary events.

Evolutionists have found that the amount of variation observed in the evolution of DNA and proteins is greater than is expected from a stochastic clock—in other words, the clock is "overdispersed," or somewhat erratic. The discrepancies in evolutionary rates along different lineages are not excessively

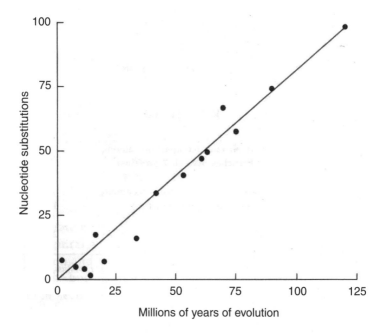

The molecular clock of evolution. The numbers of nucleotide substitutions for seven proteins in seventeen species of mammals have been estimated for each comparison between pairs of species whose ancestors diverged at the time indicated in the abscissa. Each dot represents the number of substitutions for the seven proteins added up. The line has been drawn from the origin to the outermost point and corresponds to a rate of 0.41 nucleotide substitutions per million years for all seven proteins combined. The proteins are cytochrome c, fibrinopeptides A and B, hemoglobins α and β, myoglobin, and insulin c-peptide.

large, however. So it is possible, in principle, to time phylogenetic events with considerable accuracy, but more genes or proteins (about two to four times as many) must be examined than would be required if the clock were stochastically constant, in order to achieve a desired degree of accuracy. The average rates obtained for several proteins taken together become a fairly precise clock, particularly when many species are studied.

This conclusion is illustrated in the figure on page 134 which plots the cumulative number of nucleotide changes in seven proteins against the dates of divergence of seventeen species of mammals (sixteen pairings) as determined from the fossil record. The overall rate of nucleotide substitution is fairly uniform. Some primate species (represented by the points below the line at the lower left of the figure) appear to have evolved at a slower rate than the average for the rest of the species. This anomaly occurs because the more recent the divergence of any two species, the more likely it is that the changes observed will depart from the average evolutionary rate. As the length of time increases, periods of rapid and slow evolution in any lineage are likely to cancel one another out.

The previous chapters have explored the theory of natural selection as explanation of the design of organisms and have summarized the evidence for biological evolution. In Chapter 8, I turn to issues raised by modern proponents of intelligent design. I argue that intelligent design (1) lacks scientific cogency and (2) lacks religious merit since it implies unacceptable attributes in the Creator. This discussion is followed in Chapter 9 with consideration of the tremendous power, as well as the limits, of science.

8

FOLLIES AND FATAL FLAWS

Men never do evil so completely and cheerfully as when they do it from religious conviction.

Blaise Pascal, *Pensées*, p. 894

William Paley's *Natural Theology* conveys a two-tined argument. The first prong asserts that humans, as well as all sorts of organisms, in their wholes, in their parts, and in their relations to one another and to their environment, appear to have been designed for serving certain functions and for certain ways of life. The second prong of the argument is that only an omnipotent Creator could account for the perfection and functional design of living organisms. Five and a half centuries earlier, Aquinas had been equally explicit: Only a Divine Intelligence could account for the purposefulness of the universe, which can only come from "some entity endowed with knowledge and intelligence.... Therefore some intelligent being exists by whom all natural things are directed to their end; and this being we call God."[1] Similarly, in 1691, John Ray, in *The Wisdom of God Manifested in the Works of Creation*, concluded that the functional design of all works in nature evinces that they are "Works created by God at first, and by him conserved to this Day."

Through a Glass Darkly

In the 1990s, several authors, notably biochemist Michael Behe,[2] sociologist William Dembski,[3] and law professor Phillip Johnson,[4] among others, revived the argument from design. Often, however, these authors sought to hide their real agenda, and, thus, typically avoid explicit reference to God, so that the "theory" of intelligent design (ID) could be taught in the public schools, as an alternative to the theory of evolution, without incurring conflict with the U.S. Constitution, which forbids the endorsement of any religious beliefs in public institutions.

The folly of their pretense, namely that the ID argument is scientific rather than religious, is apparent to anyone who takes the time to consider the issue seriously. As Judge John E. Jones III titles one section of the *Dover* decision: "An Objective Observer Would Know That ID and Teaching About 'Gaps' and 'Problems' in Evolutionary Theory are Creationist, Religious Strategies That Evolved from Earlier Forms of Creationism."[5] Judge Jones writes, "Although proponents of IDM [intelligent design movement] occasionally suggest that the designer could be a space alien or a time-traveling cell biologist, no serious alternative to God as a designer has been proposed by members of the IDM." Further: Professor Behe's "testimony at trial indicated that ID is only a scientific, as opposed to a religious, project for him; however, considerable evidence was introduced to refute this claim . . . ID's religious nature is evident because it involves a supernatural designer . . . expert witness ID proponents confirmed that the existence of a supernatural designer is a hallmark of ID."[6]

The duplicity of ID proponents concerning their religious objectives is particularly distressing, precisely because it is ostensibly adopted to further religion. "It is ironic," writes Judge Jones, "that several . . . individuals, who so staunchly and proudly touted their religious convictions in public, would time and again lie to cover their tracks and disguise the real purpose behind the ID policy."[7] Acknowledging that ID is a religious argument does not make it invalid. It is to this issue of the cogency of ID that I now turn by considering various claims of ID proponents.

Evolution Is Only a Theory

Opponents to teaching the theory of evolution declare that it is "only" a theory and not a fact. Indeed, they add, science relies on observation, replication, and experiment, but no one has seen the origin of the universe or the evolution of species, nor have these events been replicated in the laboratory or by experiment. These claims arise from a fundamental misunderstanding of the nature of science and how scientific theories are tested and validated.

When scientists talk about the "theory" of evolution, they use the word differently than people do in ordinary speech. In everyday speech, theory often means "guess" or "hunch," as in "I have a theory as to where Osama bin Laden is hiding." In science, however, a theory is a well-substantiated explanation of some aspect of the natural world that incorporates observations, facts, laws, inferences, and tested hypotheses. Although scientists sometimes use the word theory more casually for tentative explanations that lack substantial supporting evidence, such tentative explanations are more accurately termed "hypotheses."

According to the theory of evolution, organisms are related by common descent. There is a multiplicity of species because organisms change from generation to generation, and different lineages change in different ways. Species that share a recent ancestor are therefore more similar than those that only share remote ancestors. Thus, humans and chimpanzees are, in configuration and genetic makeup, more similar to each other than they are to baboons or to elephants.

Scientists agree that the evolutionary origin of animals and plants is a scientific conclusion beyond reasonable doubt. They place it beside such established concepts as the roundness of the Earth, its revolution around the sun, and the molecular composition of matter. That evolution has occurred is, in ordinary language, a fact.

How is this factual claim compatible with the accepted view that science relies on observation, replication, and experimentation, if no one has observed the evolution of species, much less replicated it by experiment?

What scientists observe are not the concepts or general conclusions of theories, but their consequences. Copernicus's heliocentric theory affirms that Earth revolves around the sun. Even if no one had observed this phenomenon, we accept it because of numerous confirmations of its predicted consequences.

We accept that matter is made of atoms, even if no one has seen them, because of corroborating observations and experiments in physics and chemistry.[8] The same is true of the theory of evolution. For example, the claim that humans and chimpanzees are more closely related to each other than they are to baboons leads to the prediction that the DNA of humans and chimps is more similar than that of chimps and baboons. To test this prediction, scientists select a particular gene, examine its DNA structure in each species, and thus corroborate the inference. Experiments of this kind are replicated in a variety of ways to gain further confidence in the conclusion. And so it is for myriad predictions and inferences between all sorts of organisms.

The theory of evolution makes statements about three different, though related, issues: (1) the fact of evolution, that is, that organisms are related by common descent; (2) evolutionary history—the details of when lineages split from one another and of the changes that occurred in each lineage; and (3) the mechanisms or processes by which evolutionary change occurs.

The first issue is the most fundamental and the one established with utmost certainty. Darwin gathered much evidence in its support, but evidence has accumulated continuously ever since, derived from all biological disciplines. The evolutionary origin of organisms is today a scientific conclusion established beyond reasonable doubt, endowed with the kind of certainty that scientists attribute to well-established scientific theories in physics, astronomy, chemistry, and molecular biology. This degree of certainty beyond reasonable doubt is what is implied when biologists say that evolution is a "fact"; the evolutionary origin of organisms is accepted by virtually every biologist.

The theory of evolution goes far beyond the general affirmation that organisms evolve. The second and third issues— seeking to ascertain evolutionary history, as well as to explain how and why evolution takes place—are matters of active scientific investigation. Some conclusions are well established. One, for example, is that chimpanzees are more closely related to humans than is either of those two species to baboons or to other monkeys, as mentioned above. Another conclusion is that natural selection, the process postulated by Darwin, explains the configuration of such adaptive features as the human eye and the wings of birds. Many matters are less certain, others are conjectural, and still others—such as the characteristics of the first living things and the precise time when they came about—remain largely unknown.

However, uncertainty about these issues does not cast doubt on the fact of evolution. Similarly, we do not know all the details about the configuration of the universe and the origin of the galaxies, but this is not a reason to doubt that the galaxies exist or to throw out all we have learned about their characteristics. Evolutionary biology is one of the most active fields of scientific research at present, and significant discoveries continually accumulate, supported in great part by advances in other biological disciplines.

The theory of evolution needs to be taught in the schools because nothing in biology makes sense without it. Modern biology has broken the genetic code, developed highly productive crops, and provided knowledge for improved health care. Students need to be properly trained in biology in order to improve their education, increase their chances for gainful employment, and enjoy a meaningful life in a technological world.

Learning about evolution also has practical value. The theory of evolution has made important contributions to society. Evolution explains why many human pathogens have developed resistance to formerly effective drugs and suggests ways of confronting this increasingly serious health problem. Evolutionary biology has contributed importantly to agriculture by explaining the relationships between wild and domesticated plants and between animals and their natural enemies. An understanding of evolution is indispensable for establishing sustainable relationships with the natural environment.

The Two-Explanations Fallacy

One of ID proponents' delusional assertions states, implicitly or explicitly, that if evolution fails to explain some biological phenomenon, ID must be the correct explanation. This is a misunderstanding of the scientific process. If one explanation fails, it does not necessarily follow that some other particular explanation is correct. Explanations must stand on their own evidence, not on the failure of their alternatives. Scientific explanations or hypotheses are creations of the mind, conjectures, imaginative exploits about the makeup and operation of the natural world. It is the imaginative preconception of what might be true in a particular case that guides observations and experiments designed to test whether a hypothesis is correct. The degree of acceptance of a hypothesis is related to the severity of the tests that it has passed.

It is not sufficient for a theory to be accepted because some alternative theory has failed. Oxygen was not discovered simply because it was shown that phlogiston does not exist.[9] Nor is the periodic table of chemical elements accepted just because chemical substances react and yield a variety of components. Similarly, Darwin's theory of evolution by natural selection became generally accepted by scientists not because other evolutionary theories, such as Lamarck's, Bergson's, or Darwin's grandfather Erasmus', have failed the tests of science, but because it has sustained innumerable tests and has been fertile in yielding new knowledge.

This point was forcefully made by Judge Jones in *Dover*: "ID is at bottom premised upon a false dichotomy, namely that to the extent evolutionary theory is discredited, ID is confirmed. . . . The same argument . . . was employed by creationists in the 1980s to support 'creation science.' . . . The court in *McLean* [the Arkansas federal district decision of January 5, 1982] noted the 'fallacious pedagogy of the two model approach' and that . . . 'in support of creation science, the defendants relied upon the same false premise . . . all evidence which criticized evolutionary theory was proof in support of creation science.' We do not find this false dichotomy any more availing to justify ID today than it was to justify creation science two decades ago."[10]

The Eye of the Octopus

ID proponents call for an Intelligent Designer to explain the supposed irreducible complexity in organisms. An irreducibly complex system is defined by Behe as an entity "composed of several well-matched, interacting parts that contribute to the basic function, wherein the removal of any one of the parts causes the system to effectively cease functioning."[11]

ID proponents have argued that irreducibly complex systems cannot be the outcome of evolution. According to Behe, "An irreducibly complex system cannot be produced directly . . . by slight, successive modifications of a precursor system, because any precursor to an irreducible complex system that is missing a part is by definition nonfunctional. . . . Since natural selection can only choose systems that are already working, then if a biological system cannot be produced gradually it would have to arise as an integrated unit, in one fell swoop, for natural selection to have anything to act on."[12]

In other words, unless all parts of the eye come simultaneously into existence, the eye cannot function; it does not benefit a precursor organism to have just a retina, or a lens, if the other parts are lacking. The human eye, according to this argument, could not have evolved one small step at a time, in the piecemeal manner by which natural selection works. But evolutionists have pointed out, again and again, with supporting evidence, that organs and other components of living beings are not irreducibly complex—they do not come about suddenly, or in one fell swoop. Evolutionists have shown that the organs and systems claimed by ID proponents to be irreducibly complex—such as the human eye, the biochemical mechanism of blood clotting, or the molecular rotary motor, called the flagellum, by which bacterial cells move—are not irreducible at all; rather, less complex versions of the same systems have existed in the past and can be found in today's organisms.

The human eye did not appear suddenly in all its present complexity. Its formation required the integration of many genetic units, each improving the performance of preexisting, functionally less perfect eyes. About 700 million years ago, the ancestors of today's vertebrates already had organs sensitive to light. Mere perception of light—and, later, various levels of visual ability—were beneficial to these organisms living in environments pervaded by sunlight. Different kinds of eyes that exhibit a full range of complexities and patterns have independently evolved at least forty times, although at least one gene (known as *Pax6*) is ancient and has played a role in the evolution of many different sorts of eyes.

Eyes evolved gradually and achieved very different configurations in different organisms, all serving the function of vision. Because sunlight is a pervasive feature of Earth's environment, it is not surprising that organs have evolved that take advantage of it. The simplest "organ" of vision occurs in some single-celled aquatic organisms that have enzymes or spots sensitive to light, which helps them move toward the surface of their pond, where they feed on the algae growing there. Some multicellular animals exhibit light-sensitive spots on their epidermis. Further steps—deposition of pigment around the spot, configuration of cells into a cuplike shape, thickening of the epidermis leading to the development of a lens, development of muscles to move the eyes and nerves to transmit optical signals to the brain—gradually led to the highly developed eyes of vertebrates and cephalopods (octopuses and squids) and to the compound eyes of insects.

The gradual process of natural selection, adapting organs to functions, occurs in a variety of ways that reflect the haphazard component of evolution due to mutation, past history, and the vagaries of environments. In some cases the changes of an organ involve a functional shift. An example is the evolution of the forelimbs of vertebrates, originally adapted for walking, which are used by birds for flying, by whales for swimming, and by humans for handling objects. Other cases, as with the evolution of eyes, exemplify gradual advancement of the same function—seeing. In all cases, however, the process is impelled by natural selection's favoring through time individuals that exhibit functional advantages over others of the same species.

Some transitions at first may seem unlikely because it is often difficult to identify which possible functions may have

---

Facing page: Eyes in living mollusks. The octopus eye (bottom) is quite complex, with components similar to those of the human eye, such as cornea, iris, refractive lens, and retina. Other mollusks have simpler eyes. The simplest eye is found in limpets (top), consisting of only a few pigmented cells, slightly modified from typical epithelial (skin) cells. Slitshell mollusks (second from top) have a slightly more advanced organ, consisting of some pigmented cells shaped as a cup. Further elaborations and increasing complexity are found in the eyes of *Nautilus* and *Murex*, not yet as complex as the eyes of the squid and octopus. (Adapted from "Evolution, The Theory of," courtesy of Encyclopaedia Britannica, Inc.)

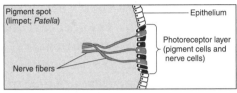

Pigment spot (limpet; *Patella*)

Nerve fibers

Epithelium

Photoreceptor layer (pigment cells and nerve cells)

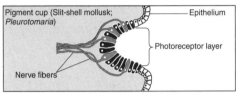

Pigment cup (Slit-shell mollusk; *Pleurotomaria*)

Nerve fibers

Epithelium

Photoreceptor layer

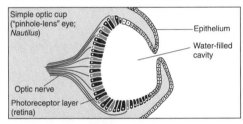

Simple optic cup ("pinhole-lens" eye; *Nautilus*)

Optic nerve

Photoreceptor layer (retina)

Epithelium

Water-filled cavity

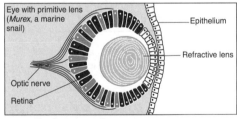

Eye with primitive lens (*Murex*, a marine snail)

Optic nerve

Retina

Epithelium

Refractive lens

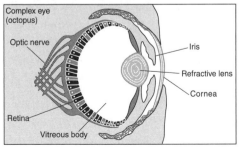

Complex eye (octopus)

Optic nerve

Retina

Vitreous body

Iris

Refractive lens

Cornea

been served during the intermediate stages; but these cases are eventually resolved with further research and the discovery of intermediate fossil forms or living organisms with intermediate stages of development, as in the case of mollusks' eyes. An example of a seemingly unlikely transition is described above, namely, the transformation of bones found in the reptilian jaw into the hammer and anvil of the mammalian ear. I now turn to some examples given by ID leaders.

The Bacterial Flagellum, Blood Clotting, and Other Improbabilities

One favorite ID example of alleged irreducible complexity is the bacterial flagellum. The bacterial flagellum is, according to Behe, irreducibly complex because it consists of several parts so that, if any part is missing, the flagellum will not function. It could not, therefore, says Behe, have evolved gradually, one part at a time, because the function belongs to the whole; the separate parts cannot function by themselves. "Because the bacterial flagellum is necessarily composed of at least three parts—a paddle, a rotor, and a motor—it is irreducibly complex."[13] This inference is, of course, incorrect.

The flagellum is embedded in the cell membrane of the bacteria. The external swimming element that functions as paddle, or propeller, is a filament consisting of a single kind of protein, called "flagellin." At the cell membrane, the filament attaches to a "rotor," made up of a so-called "hook protein." The motor that rotates the filament is located at the base of the flagellum and consists of two elements: a rotor (the part that rotates) and a stator (the stationary component).

The argument that the different components of the flagellum must have come about "in one fell swoop," because the parts cannot function separately and thus could not have evolved independently, is reminiscent of Paley's argument about the eye. Of what possible use would be the iris, cornea, lens, retina, and optic nerve, one without the others? Yet we know that component elements of the octopus eye can evolve gradually, cumulatively, and that simple eyes, as they exist in a limpet, in shell mollusks, and in marine snails are functional.

The bacterial flagellum does not exist. In different species of bacteria, there are different kinds of flagella, some simpler than the one described by Behe, others just different, even very different, as in the archaea, a very large group of bacteria-like organisms. Twenty of the 42 proteins of the *Salmonella typhimurium* flagellum (the model case described by Behe) are required in the flagella of other bacterial species.[14] Moreover, motility in many bacteria is accomplished without flagella at all. Further, biochemists have shown that some flagellum components may have evolved from secretory systems, which are very similar to the flagellum, but lack some of the flagellum's components. A major component of the flagellum described by Behe has essentially the same structure as type-III secretory systems (TTSS), although these lack the motor proteins.

There are many kinds of disease-causing bacteria (although not all bacteria, or even a majority, are agents of human disease). One way in which bacteria cause disease is by injecting toxins (poisons) into the cells of the host organism, which they accomplish by means of special protein secretory systems, one of which is TTSS. It turns out that the TTSS proteins and portions of the bacterial flagellum are homologous, that is, are very similar and have a common evolutionary origin. The bacterial flagellum is not irreducibly complex: A subset of the flagellum's complement of proteins evolved as a mechanism for bacteria to inject proteins across a cell's membrane. Of the 20 proteins of the *Salmonella typhimurium* flagellum universally required in other bacterial flagella, 18 have homologous related proteins that function in other, simpler biochemical systems.[15]

The argument for the irreducible complexity of the flagellum is formulated, like other ID arguments, as an "argument from ignorance." Because one author (Behe, in this case) does not know how a complex organ may have come about, it must be the case that it is irreducibly complex. This argument from ignorance dissolves as scientific knowledge advances, or when preexisting scientific knowledge is taken into account.

This book is hardly the appropriate venue to enter into exhaustive technical details of how bacterial flagella may have evolved gradually, derived from structures originally evolved for different functions, such as secretion, nor is it the place to cite the scientific papers where the technical details are given. General discussions can be found in the writings, for example, of Ian Musgrave,[16] David Ussery,[17] and Kenneth Miller,[18] who cite the original scientific literature. In reviewing the matter, Miller points out that "The most powerful rebuttals of the flagellum story, however, have not come from direct attempts to answer the critics of evolution. Rather, they have emerged from the steady progress of scientific work on the genes and proteins associated with the flagellum and other cellular structures. Such studies have now established that the entire premise by which this molecular machine has been advanced as an argument against evolution is wrong—the bacterial flagellum is not irreducibly complex."[19]

Let's consider another ID favorite, blood clotting. An injured person bleeds for a short time until a clot forms, which soon hardens and the bleeding stops. As Behe writes, "Blood clotting is a very complex, intricately woven system consisting of a score of interdependent protein parts."[20] Blood clotting is such a complex process, and seemingly so unnecessarily complex, that Behe has compared it to a machine designed by Rube Goldberg, the great cartoonist who designed very complex machines to perform tasks that could be accomplished much more simply. The coagulating mechanism is one of Behe's examples of intelligently designed biochemical processes. According to Behe, "*No one on earth has the vaguest idea how the coagulation cascade came to be*" (his italics).[21]

This is a remarkable statement, particularly because of the numerous scientific papers about the evolution of the various components of the blood-clotting mechanism in vertebrates, including "The Evolution of Vertebrate Blood Coagulation: A Case of Yin and Yang" by the eminent biochemist Russell F. Doolittle, published in 1993,[22] 3 years before Behe's *Darwin's Black Box*. The biochemical and genetic components of the process are excessively complex to be described here in detail. Non-technical summaries can be found in Kenneth R. Miller, *Finding Darwin's God*, pages 152–161, as well as in other publications.[23] The clotting process involves a soluble protein called fibrinogen, which makes up about 3 percent of the protein in the blood plasma of vertebrates. Fibrinogen has a sticky portion near the center of the molecule, which is covered by short amino acid chains, which, when bleeding begins are clipped off by a protease enzyme (thrombin), so that the remaining fibrins stick together and begin the formation of a clot. The activation of thrombin requires additional proteases in a cascade that involves five or six steps, by which the biochemical signal becomes amplified more than a million times, so that the clot forms rapidly. The genes coding for the various proteases have evolved by several gene duplications over evolutionary time. These gene duplications have been reconstructed in human ancestry and, independently, in other vertebrates. Simpler clotting mechanisms occur in crabs, lobsters, starfish, sea urchins, worms, and other animals with circulatory systems. The clotting mechanism of vertebrates evolved from simpler mechanisms similar to those now found in invertebrates, as shown by reconstruction of the evolution of the genes involved.

Astonishingly, Michael Behe has written that "There is no publication in the scientific literature—in prestigious journals, specialty journals, or books—that describes how molecular evolution of any real, complex, biochemical system either did occur or even might have occurred" and, in particular, "the scientific literature has no answers to the origin of the immune system."[24] In *Dover*, Judge Jones points out with understated disbelief that "Professor Behe was questioned concerning his 1996 claim that science would never find an evolutionary explanation for the immune system. He was presented with fifty-eight peer-reviewed publications, nine books, and several immunology textbook chapters about the evolution of the immune system; however, he simply insisted that this was still not sufficient evidence of evolution, and that it was not 'good enough.'" Judge Jones concludes: "We therefore find that Professor Behe's claim for irreducible complexity has been refuted in peer-reviewed research papers and has been rejected by the scientific community at large."[25]

Darwin wrote, as Behe has quoted, "If it could be demonstrated that any complex organ existed, which could not possibly have been formed by numerous, successive, slight modifications, my theory would absolutely break down." Behe fails to quote the ensuing sentence: "But I can find no such case." Scientific advances allow us to reiterate Darwin's conclusion, even more forcefully than he did. Neither Behe, nor any other IDer or anybody else, has found any instance of an organ or complex process that could not be explained through gradual evolution of its components, even though not all the details may be known.

Theodosius Dobzhansky, one of the greatest evolutionists of the twentieth century and a religious person, has written: "There are people, however, to whom the gaps in our understanding of nature are pleasing for a different reason. These people hope that the gaps will be permanent, and that what is unexplained will also remain inexplicable. By a curious twist of reasoning, what is unexplained is then assumed to be the realm of divine activity. The historical odds are all against the 'God of the gaps' being able to retain these shelters in perpetuity. There is nothing, however, that can satisfy the type of mind which refuses to accept this testimony of historical experience."[26]

William Dembski[27] has upstaged Behe's "discovery" of irreducible complexity by reducing it to a special case of "complex specified information," which is information that has a very low prior probability and, therefore, high information content. Dembski argues that mutation and natural selection are incapable of generating such highly improbable states of affairs. Take the 30 proteins that make up the bacterial flagellum. Assume that each protein has about 300 amino acids, he calculates that the probability of one such protein is 20^{-300}. After some refinements, he calculates that the probability of origination for the flagellum is 10^{-1170} (one divided by 1 followed by 1170 zeroes; in order to get some idea of the magnitude of this number, consider that the number of atoms in the universe is estimated to be 10^{77}, or 1 followed by 77 zeroes). Dembski concludes that even if one would take into account that life has existed on Earth for 3.5 billion years, the assembly of a functioning flagellum is impossibly improbable.

What are we to make of this calculation? The answer is simple, namely, that this calculation, as well as Dembski's other numerology exercises, is totally irrelevant because Dembski's assumptions are wrong. Natural selection proceeding stepwise can accomplish outcomes with prior probabilities immensely smaller than Dembski's calculations. The explanation can be found in Chapter 4. In the example given, the probability that one single bacterium would acquire resistance to the antibiotic streptomycin and the ability to synthesize the amino acid histidine is 4×10^{-16} (four in ten thousand trillion) Yet all 20 to 30 billion bacteria in the final culture exhibit these properties. This adaptation has been accomplished by natural selection in response to environmental changes in just a few days.

In Praise of Imperfection

One difficulty with attributing the design of organisms to the Creator is that imperfections and defects pervade the living world. Consider the human eye. The visual nerve fibers in the eye converge to form the optic nerve, which crosses the retina (in order to reach the brain) and thus creates a blind spot, a minor imperfection, but an imperfection of design, nevertheless; squids and octopuses do not have this defect. Did the Designer have greater love for squids than for humans and, thus, exhibit greater care in designing their eyes than ours?

Defective design would seem incompatible with an omnipotent Intelligent Designer. Anticipating this criticism, Paley responded that "apparent blemishes . . . ought to be referred to some cause, though we be ignorant of it." Modern ID proponents have made similar assertions. According to Behe, "The argument from imperfection overlooks the possibility that the designer might have multiple motives, with engineering excellence oftentimes relegated to a secondary role. . . . [T]he reasons that a designer would or would not do anything are virtually impossible to know unless the designer tells you specifically what those reasons are."[28] This statement, scientists and philosophers have responded, may have theological validity, but it destroys ID as a scientific hypothesis, because it provides ID with an empirically impenetrable shield against predictions. Since we know not how "intelligent" or "perfect" a design will be, whether it has been intelligently designed cannot be empirically tested.[29] Science tests its hypotheses by observing whether or not predictions derived from them hold true in the observable world. A hypothesis that cannot be tested empirically—that is, by observation or experiment—is not scientific. ID as an explanation for the adaptations of organisms could be (natural) theology, as Paley would have it, but, whatever it is, it is not a scientific hypothesis.

The theory of ID is not good theology either because it leads to conclusions about the nature of the Designer quite different from those of omniscience, omnipotence, and benevolence that Paley had inferred as the attributes of the Creator and that Christian theology predicates of God. It is not only that organisms and their parts are less than perfect, but also that deficiencies and dysfunctions are pervasive, evidencing "incompetent" rather than "intelligent" design. Consider the human jaw. We have too many teeth for the jaw's size, so that wisdom teeth need to be removed and orthodontists can make a decent living straightening the others. Would we want to blame God for this blunder? A human engineer would have done better.

Evolution gives a good account of this imperfection. Brain size increased over time in our ancestors; the remodeling of the skull to fit the larger brain entailed a reduction of the jaw, so that the head of the newborn would not be too large to pass through the mother's birth canal. Evolution responds to the organisms' needs through natural selection, not by optimal design but by "tinkering," by slowly modifying existing structures. Evolution achieves "design," as a consequence of natural selection while promoting adaptation. Evolution's ID is *imperfect* design, not *intelligent* design.

Consider the birth canal of women, much too narrow for easy passage of the infant's head, so that thousands upon thousands of babies and many mothers die during delivery. Surely we don't want to blame God for this dysfunctional design or for the children's deaths. Science makes it understandable, a consequence of the evolutionary enlargement of our brain. Females of other primates do not experience this difficulty. Theologians in the past struggled with the issue of dysfunction because they thought it had to be attributed to God's design. Science, much to the relief of theologians, provides an explanation that convincingly attributes defects, deformities, and dysfunctions to natural causes.

Another human example: Why are our arms and our legs, which are used for such different functions, made of the same materials, the same bones, muscles, and nerves, all arranged in the same overall pattern? Evolution makes sense of the anomaly. Our remote ancestors' forelimbs were legs. After our hominid ancestors became bipedal and started using their forelimbs for functions other than walking, the forelimbs became gradually modified, but retaining their original composition and arrangement. Engineers start with raw materials and a design suited for a particular purpose; evolution can only modify what is already there. An engineer who would design cars and airplanes, or wings and wheels, using the same materials arranged in a similar pattern, would surely be fired.

More disturbing yet for ID proponents has to be the following consideration. About 20 percent of all recognized human pregnancies end in spontaneous miscarriage during the first two months of pregnancy. This misfortune amounts at present to more than 20 million spontaneous abortions worldwide every year. Do we want to blame God for the deficiencies in the pregnancy process? Is God the greatest abortionist of them all? Most of us might rather attribute this monumental mishap to the clumsy ways of the evolutionary process than to the incompetence of an intelligent designer.

Examples of deficiencies and dysfunctions in all sorts of organisms can be listed endlessly, reflecting the opportunistic, tinkerer-like character of natural selection, which achieves imperfect, rather than intelligent, design. The world of organisms also abounds in characteristics that might be called "oddities," as well as those that have been characterized as "cruelties," an apposite qualifier if the cruel behaviors were designed outcomes of a being holding onto human or higher standards of morality. However, the cruelties of biological nature are only metaphoric cruelties when applied to the outcomes of natural selection.

Examples of "cruelty" involve not only the familiar predators tearing apart their prey (say, a small monkey held alive by a chimpanzee biting large flesh morsels from the screaming monkey), or parasites destroying the functional organs of their hosts, but also, and very abundantly, between organisms of the same species, even between mates. A well-known example is the female praying mantis that devours the male after coitus is completed. Less familiar is that, if she gets the opportunity, the female praying mantis will eat the head of the male *before* mating, which thrashes the headless male mantis into spasms of "sexual frenzy" that allow the female to connect his genitalia with hers.[30] In some midges (tiny flies), the female captures the male as if he were any other prey and with the tip of her proboscis she injects into him her spittle, which starts digesting the male's innards which are then sucked by the female; partly protected from digestion are the relatively intact male organs that break off inside the female and fertilize her.[31] Male cannibalism by their female mates is known in dozens of species, particularly spiders and scorpions.[32] The world of life abounds in "cruel" behaviors: numerous predators eat their prey alive; parasites destroy their living hosts from within; and, as noted, females of many species of spiders and insects devour their mates.

In a letter to Joseph Hooker, Darwin rued: "What a book a Devil's Chaplain might write on the clumsy, wasteful, blundering low & horridly cruel works of nature." Darwin repeatedly returned to this theme, particularly in his extensive correspondence with the American biologist Asa Gray. In 1860, he wrote to Gray that "I cannot persuade myself that a beneficent God would have designedly created the *Ichneumonidae* with the express intention of their feeding within the living bodies of caterpillars, or that a cat should play with mice."[33]

The design of organisms is often so dysfunctional, odd, and cruel that it possibly might be attributed to the gods of the ancient Greeks, Romans, and Egyptians, who fought with one another, made blunders, and were clumsy in their endeavors. For a modern biologist who knows about the world of life, the design of organisms is not compatible with special action by the omniscient and omnipotent God of Judaism, Christianity, and Islam. The American philosopher David Hull has made the same point with shrill language:

> What kind of God can one infer from the sort of phenomena epitomized by the species on Darwin's Galápagos Islands? The evolutionary process is rife with happenstance, contingency,

incredible waste, death, pain and horror. . . . Whatever the God implied by evolutionary theory and the data of natural selection may be like, he is not the Protestant God of waste not, want not. He is also not the loving God who cares about his productions. He is not even the awful God pictured in the Book of Job. The God of the Galápagos is careless, wasteful, indifferent, almost diabolical. He is certainly not the sort of God to whom anyone would be inclined to pray.[34]

The God of love and mercy could not have planned all this.

Religious scholars in the past had struggled with imperfection, dysfunction, and cruelty in the living world, which are difficult to explain if they are the outcome of God's design. The philosopher David Hume set the problem succinctly with brutal directedness: "Is he [God] willing to prevent evil, but not able? Then he is impotent. Is he able, but not willing? Then he is malevolent. Is he both able and willing? Whence then evil?"[35] Evolution came to the rescue. Jack Haught, a contemporary Roman Catholic theologian, has written of "Darwin's gift to theology."[36] The Protestant theologian Arthur Peacocke has referred to Darwin as the "disguised friend," by quoting the earlier theologian Aubrey Moore, who in 1891 wrote that "Darwinism appeared, and, under the guise of a foe, did the work of a friend."[37] Haught and Peacocke are acknowledging the irony that the theory of evolution, which at first had seemed to remove the need for God in the world, now has convincingly removed the need to explain the world's imperfections as failed outcomes of God's design.[38]

Indeed, a major burden was removed from the shoulders of believers when convincing evidence was advanced that the design of organisms need not be attributed to the immediate agency of the Creator, but rather is an outcome of natural processes. If we claim that organisms and their parts have been specifically designed by God, we have to account for the incompetent design of the human jaw, the narrowness of the birth canal, and our poorly designed backbone, less than fittingly suited for walking upright. Proponents of ID would do well to acknowledge Darwin's revolution and accept natural selection as the process that accounts for the design of organisms, as well as for the dysfunctions, oddities, cruelties, and sadism that pervade the world of life. Attributing these to specific agency by the Creator amounts to blasphemy. Proponents and followers of ID are surely well-meaning people who do not intend such blasphemy, but this is how matters appear to a biologist concerned that God not be slandered with the imputation of incompetent design.

In Chapter 9, I seek to complete the circle of my argumentation by making the point that religion and science do not stand in opposition because they concern different realms of reality. Rather, they may be seen as complementary. Questions about the meaning and purpose of the world and of human life transcend science. Religion answers them.

9

BEYOND BIOLOGY

Many of the leading proponents of ID make a bedrock assumption which is utterly false. Their presupposition is that evolutionary theory is antithetical to a belief in the existence of a supreme being and to religion in general.

Judge John E. Jones
Kitzmiller v. Dover Area School District, p. 136

In the sentence following the quoted text above, Judge Jones adds: "Repeatedly in this trial, . . . scientific experts testified that the theory of evolution represents good science, is overwhelmingly accepted by the scientific community, and that in no way conflicts with, nor does it deny, the existence of a divine creator."

It is this matter of the intelligent design (ID) proponents' perceived contradiction between the theory of evolution and a Divine Creator that I take up in this chapter. This presupposition, I want to note at the outset, is shared between ID proponents and materialistic scientists and philosophers. Strange bedfellows, indeed.

I will briefly review the history of Christianity's response to the theory of evolution and the history of the creationism movement in the United States, and then consider the powers and limits of science. We will see that scientific knowledge and

religious belief need not be in contradiction. If they are correctly assessed, they *cannot* be in contradiction, because science
and religion concern non-overlapping realms of knowledge.
It is only when assertions are made beyond their legitimate
boundaries that evolutionary theory and religious belief appear
to be antithetical.

On April 28, 1937, early in the Spanish Civil War, Nazi airplanes
under Franco's command bombed the small Basque town of
Guernica, the spiritual home of the Basques, killing 1,654 of its
7,000 inhabitants: the first time that a civilian population had
been determinedly destroyed from the air. The Spanish painter
Pablo Picasso had recently been commissioned by the Spanish
Republican government to paint a large composition for the
Spanish pavilion at the Paris World Exhibition of 1937. In a
frenzy of manic energy, the enraged Picasso sketched in two
days and fully outlined in ten more days his famous *Guernica*,
an immense painting measuring 25 feet, 8 inches by 11 feet,
6 inches.

Suppose that I list the coordinates of all images represented
in the painting, their size, the pigments used, and the quality of the canvas. This information would be interesting, but
it would hardly be satisfying if I completely omitted aesthetic
considerations and failed to reflect on the painting's meaning
and purpose, the dramatic message of man's inhumanity to
man conveyed by the outstretched figure of the mother pull-
ing her killed baby, the bellowing human faces, the wounded
horse, and the satanic image of the bull. The point is that the
physical description of the painting does not tell us anything
(by itself it cannot tell us anything) about the aesthetic value or
historical significance of *Guernica*; nor, on the other hand, do

aesthetics or intended meaning determine the physical features of the painting.

Let *Guernica* be a metaphor of the point I wish to make. Scientific knowledge, like the description of the size, materials, and geometry of *Guernica*, is satisfying and useful, but once science has had its say, there remains much about reality that is of interest: questions of value, meaning, and purpose that are forever beyond science's scope.

Evolution and the Bible

To some people, the theory of evolution seems to be incompatible with their religious beliefs, particularly those of Christians, because it is inconsistent with the Bible's narrative of creation. The first chapters of the biblical book of Genesis describe God's creation of the world, plants, animals, and human beings. A literal interpretation of Genesis seems incompatible with the gradual evolution of humans and other organisms by natural processes. Even independent of the biblical narrative, the Christian beliefs in the immortality of the soul and in humans as "created in the image of God" have appeared to many as contrary to the evolutionary origin of humans from nonhuman animals.

Religiously motivated attacks against the theory of evolution started during Darwin's lifetime. In 1874, Charles Hodge, an American Protestant theologian, published *What Is Darwinism?*, one of the most articulate assaults on evolutionary theory. Hodge perceived Darwin's theory as "the most thoroughly naturalistic that can be imagined and far more atheistic than that of his predecessor Lamarck." Echoing Paley, Hodge argued that the design of the human eye reveals that "it has been planned by the Creator, like the design of a watch evinces a watchmaker." He concluded that "the denial of design in nature is actually the denial of God."[1]

Other Protestant theologians saw a solution to the apparent contradiction between evolution and creation in the argument that God operates through intermediate causes. The origin and motion of the planets could be explained by the law of gravity and other natural processes without denying God's creation and providence. Similarly, evolution could be seen as the natural process through which God brought living beings into existence and developed them according to his plan. Thus, A. H. Strong, the president of Rochester Theological Seminary in New York State, wrote in his *Systematic Theology* (1885): "We grant the principle of evolution, but we regard it as only the method of divine intelligence." He explains that the brutish ancestry of human beings was not incompatible with their excelling status as creatures in the image of God. Strong drew an analogy with Christ's miraculous conversion of water into wine: "The wine in the miracle was not water because water had been used in the making of it, nor is man a brute because the brute has made some contributions to its creation."[2] Arguments for and against Darwin's theory came from Roman Catholic theologians as well.

Gradually, well into the twentieth century, evolution by natural selection came to be accepted by a majority of Christian writers. Pope Pius XII in his encyclical *Humani generis* (1950, Of the Human Race) acknowledged that biological evolution was compatible with the Christian faith, although he argued that God's intervention was necessary for the creation of the human soul. Pope John Paul II, in an address to the Pontifical Academy of Sciences on October 22, 1996, deplored interpreting the Bible's texts as scientific statements rather than religious teachings. He added: "New scientific knowledge has led us to realize that the theory of evolution is no longer a mere hypothesis. It is indeed remarkable that this theory has been progressively accepted by researchers, following a series of discoveries in various fields of knowledge. The convergence, neither sought nor fabricated, of the results of work that was conducted independently is in itself a significant argument in favor of this theory."[3]

Similar views have been expressed by other mainstream Christian denominations. The General Assembly of the United Presbyterian Church in 1982 adopted a resolution stating that "Biblical scholars and theological schools . . . find that the scientific theory of evolution does not conflict with their interpretation of the origins of life found in Biblical literature." The Lutheran World Federation in 1965 affirmed that "evolution's assumptions are as much around us as the air we breathe and no more escapable. At the same time theology's affirmations are being made as responsibly as ever. In this sense both science and religion are here to stay, and . . . need to remain in a healthful tension of respect toward one another."[4]

Similar statements have been advanced by Jewish authorities and leaders of other major religions. In 1984, the 95th Annual Convention of the Central Conference of American Rabbis adopted a resolution stating: "Whereas the principles and concepts of biological evolution are basic to understanding science . . . we call upon science teachers and local school authorities in all states to demand quality textbooks that are based on modern, scientific knowledge and that exclude 'scientific' creationism."

Christian denominations that hold a literal interpretation of the Bible have opposed these views. A succinct expression of this opposition is found in the Statement of Belief of the Creation Research Society, founded in 1963 as a "professional organization of trained scientists and interested laypersons who are firmly committed to scientific special creation": "The Bible is the Written Word of God, and because it is inspired throughout, all of its assertions are historically and scientifically true in the original autographs. To the student of nature this means that the account of origins in Genesis is a factual presentation of simple historical truths."

Many Bible scholars and theologians have long rejected a literal interpretation as untenable, however, because the Bible contains mutually incompatible statements. The very beginning of the book of Genesis presents two different creation narratives. Extending through Chapter 1 and the first verses of Chapter 2 is the familiar six-day narrative, in which God creates human beings—both "male and female"—in his own image on the sixth day, after creating light, earth, firmament, fish, fowl, and cattle. In verse 4 of Chapter 2, a different narrative starts, in which God creates a male human, then plants a garden and creates the animals, and only then proceeds to take a rib from the man to make a woman.

Which one of the two narratives is correct and which one is in error? Neither one contradicts the other, I would say, if we understand the two narratives as conveying the same message, that the world was created by God and that humans are His creatures. But both narratives cannot be "historically and scientifically true" as postulated in the Statement of Belief of the Creation Research Society.

There are numerous inconsistencies and contradictions in different parts of the Bible, for example, in the description of the return from Egypt to the promised land by the chosen people of Israel, not to mention erroneous factual statements about the sun's circling around the Earth and the like. Biblical scholars point out that the Bible is inerrant with respect to religious truth, not in matters that are of no significance to salvation. Augustine, one of the greatest Christian theologians, wrote in his *De Genesi ad litteram* (Literal Commentary on Genesis): "It is also frequently asked what our belief must be about the form and shape of heaven, according to Sacred Scripture. . . . Such subjects are of no profit for those who seek beatitude. And what is worse, they take up very precious time that ought to be given to what is spiritually beneficial. What concern is it of mine whether heaven is like a sphere and earth is enclosed by it and suspended in the middle of the universe, or whether heaven is like a disk and the Earth is above it and hovering to one side."[5]

Augustine adds later in the same chapter: "In the matter of the shape of heaven, the sacred writers did not wish to teach men facts that could be of no avail for their salvation." Augustine is saying that the book of Genesis is not an elementary book of astronomy. Indeed, Augustine noted that in the Genesis narrative of creation, God creates light on the first day but did not create the sun until the fourth day. Augustine concluded that "light" and "days" in Genesis make no literal sense.[6] The Bible is about religion, and it is not the purpose of the Bible's religious authors to settle questions about the shape of the universe that are of no relevance whatsoever to how to seek salvation.

In the same vein, Pope John Paul II said in 1981 that the Bible itself

> speaks to us of the origins of the universe and its makeup, not in order to provide us with a scientific treatise but in order to state the correct relationships of man with God and with the universe. Sacred Scripture wishes simply to declare that the world was created by God, and in order to teach this truth, it expresses itself in the terms of the cosmology in use at the time of the writer. . . . Any other teaching about the origin and makeup of the universe is alien to the intentions

of the Bible, which does not wish to teach how heaven was
made but how one goes to heaven.[7]

John Paul's argument was clearly a response to Christian fundamentalists who see in Genesis a literal description of how the
world was created by God.

Fundamentalism, Creationism, and the Public Schools

In modern times, biblical fundamentalists, although a minor-
ity of Christians in the United States, have periodically gained
considerable public and political influence. Opposition to the
teaching of evolution can largely be traced to two movements
with nineteenth-century roots, Seventh-day Adventism and
Pentecostalism. Consistent with their emphasis on the seventh-
day Sabbath as a memorial of the biblical Creation, Seventh-
day Adventists have insisted on the recent creation of life and
the universality of the Flood, which they believe deposited the
fossil-bearing rocks. This distinctively Adventist interpretation
of Genesis became the hard core of "creation science" in the late
twentieth century and was incorporated into the "balanced-
treatment" laws of Arkansas and Louisiana (see below). Many
Pentecostals, who generally endorse a literal interpretation of
the Bible, also have adopted and endorsed the tenets of creation
science, including the recent origin of Earth and a geology
interpreted in terms of the Flood. They differ from Seventh-
day Adventists and other adherents of creation science, however, in their tolerance of diverse views and the limited import
they attribute to the evolution-creation controversy.

During the 1920s, biblical fundamentalists helped influence more than twenty state legislatures to debate antievolution
legislation, and four states—Arkansas, Mississippi, Oklahoma,
and Tennessee—prohibited the teaching of evolution in their

public schools. A spokesman for the antievolutionists was William Jennings Bryan, three times the unsuccessful Democratic candidate for the U.S. presidency, who said in 1922, "We will drive Darwinism from our schools." In 1925, Bryan took part in the prosecution of John T. Scopes, a high school teacher in Dayton, Tennessee, who had admittedly violated the state's law forbidding the teaching of evolution.

In 1968 the Supreme Court of the United States declared unconstitutional any law banning the teaching of evolution in public schools (*Epperson v. Arkansas* 393 U.S.97, 1968). Thereafter, Christian fundamentalists introduced legislation in a number of state legislatures ordering that the teaching of "evolution science" be balanced by allocating equal time to "creation science." Creation science, it was asserted, propounds that all kinds of organisms abruptly came into existence when God created the universe, that the world is only a few thousand years old, and that the biblical Flood was an actual event that only one pair of each animal species survived. The legislatures of Arkansas in 1981 and Louisiana in 1982 passed statutes requiring the balanced treatment of evolution science and creation science in their schools, but opponents successfully challenged the statutes as violations of the constitutionally mandated separation of church and state.

The Arkansas statute was declared unconstitutional in federal court in 1982 after a public trial in Little Rock.[8] The Louisiana law was appealed all the way to the Supreme Court of the United States, which in 1987 ruled Louisiana's "Creationism Act" unconstitutional because, by advancing the religious belief that a supernatural being created humankind, which is embraced by the phrase "creation science," the act impermissibly endorses religion.[9] The most recent confrontation between creationism and the theory of evolution in the courts of law involves the concept of intelligent design (ID), which in its current formulation came into existence after the Supreme Court's decision of 1987 that creation science could not be taught in the public schools.

December 2005: The *Dover* Decision

On October 28, 2004, the Dover (Pennsylvania) Area School Board of Directors adopted the following resolution: "Students will be made aware of gaps/problems in Darwin's theory and of other theories of evolution including, but not limited to, intelligent design."

Further, on November 19, 2004, the Dover Area School District announced by press release that, starting in January 2005, teachers would be required to read a statement, which includes the following assertions: "Because Darwin's Theory is a theory, it continues to be tested as new evidence is discovered. The Theory is not a fact. . . . Intelligent Design is an explanation of the origin of life that differs from Darwin's view. The reference book *Of Pandas and People*, is available for students who might be interested in gaining an understanding of what Intelligent Design actually involves."

The constitutional validity of the resolution and press release was challenged on December 14, 2004, in the Federal District Court for the Middle District of Pennsylvania by eleven parents (Kitzmiller et al.). The trial took place over several weeks in the fall of 2005. On December 20, 2005, Federal Judge John E. Jones III issued a 139-page-long decision (*Kitzmiller v. Dover Area School District*), declaring that "the Defendants' ID Policy violates the Establishment Clause of the First Amendment of the Constitution of the United States" and that the "Defendants are permanently enjoined from maintaining the ID Policy in any school within the Dover Area School District" (p. 139).

Judge Jones reviews the history of the creationist and ID movements in the United States, and affirms that "The overwhelming evidence at trial established that ID is a religious view, a mere re-labeling of creationism, and not a scientific theory" (p. 43). "ID is not supported by any peer-reviewed research, data, or publications" (p. 87). "It has not generated peer-reviewed publications, nor has it been the subject of testing and research" (p. 64). "ID is not science and cannot be adjudged a valid, accepted scientific theory" (p. 89). "In summary, the disclaimer . . . misrepresents its [the theory of evolution] status in the scientific community, causes students to doubt its validity without scientific justification, presents students with a religious alternative masquerading as a scientific theory, . . . and instructs students to forego scientific inquiry" (p. 49).

Toward the end of the decision, Judge Jones minces no words when referring to the School Board's ID supporters: "This case came to us as a result of the activism of an ill informed faction on a school board, aided by a national public interest law firm. . . . The breathtaking inanity of the Board's decision is evident when considered against the factual backdrop which has now been fully revealed through this trial. The students, parents, and teachers of the Dover Area School District deserved better" (pp. 137–138).

Whether or not *Dover* will be appealed in the courts remains to be seen. In any case, the efforts of fundamentalist creationism toward discrediting the theory of evolution will surely persist.

Evolution or Religious Beliefs? The Arrogance of Exclusivity

Does Darwinism exclude religious beliefs? Is science fundamentally materialistic? The answer to the first question is no.

The answer to the second question is: depends. It depends on whether it refers to scientific scope and methodology or to metaphysical conceits.

The scope of science is the world of nature, the reality that is observed, directly or indirectly, by our senses. Science advances explanations concerning the natural world, explanations that are subject to the possibility of corroboration or rejection by observation and experiment. Outside that world, science has no authority, no statements to make, no business whatsoever taking one position or another. Science has nothing decisive to say about values, whether economic, aesthetic, or moral; nothing to say about the meaning of life or its purpose; nothing to say about religious beliefs (except in the case of beliefs that transcend the proper scope of religion and make assertions about the natural world that contradict scientific knowledge; such statements cannot be true).

Science is *methodologically* materialistic or, better, methodologically *naturalistic*. I prefer the second expression because "materialism" often refers to a metaphysical conception of the world, a philosophy that asserts that nothing exists beyond the world of matter, that nothing exists beyond what our senses can experience. That is why I averred that the question whether or not science is inherently materialistic depends on whether we are referring to the methods and scope of science, which remain within the world of nature, or to the metaphysical implications of materialistic philosophy asserting that nothing exists beyond the world of matter. Science does not imply metaphysical materialism.

Scientists and philosophers who assert that science excludes the validity of any knowledge outside science make a "categorical mistake," confuse the method and scope of science with its metaphysical implications. Methodological naturalism asserts the boundaries of scientific knowledge, not its universality. Science transcends cultural, political, and religious differences because it has no assertions to make about these subjects (except, again, to the extent that scientific knowledge is negated). That science is not constrained by cultural or religious differences is one of its great virtues. Science does not transcend those differences by denying them or by taking one position rather than another. It transcends cultural, political, and religious differences because these matters are none of its business.

Yet, some scientists, including evolutionists, assert that science denies any valid knowledge concerning values or the world's meaning and purpose. The distinguished evolutionist and textbook author Douglas Futuyma avers: "By coupling undirected, purposeless variation to the blind, uncaring process of natural selection, Darwin made theological or spiritual explanations of the life processes superfluous. . . . Darwin's theory of evolution . . . provided a crucial plank in the platform of mechanism and materialism."[10]

The well-known evolutionary biologist Richard Dawkins explicitly denies design, purpose, and values: "the universe that we observe has precisely the properties we should expect if there is, at bottom, no design, no purpose, no evil and no good, nothing but blind, pitiless indifference."[11] The historian of science William Provine not only affirms that there are no absolute principles of any sort, but draws the ultimate conclusion from a materialistic line of thinking that even free will is an illusion: "Modern science directly implies that there are no inherent moral or ethical laws, no absolute guiding principles for human society. . . . [F]ree will as it is traditionally conceived—the freedom to make uncoerced and unpredictable choices among alternative courses of action—simply does not exist."[12]

There is a monumental contradiction in these assertions. If its commitment to naturalism does not allow science to derive values, meanings, or purposes from scientific knowledge, it surely does not allow it, either, to deny their existence. We may grant these authors their right to think as they wish, but they have no warrant whatsoever to ground their materialistic philosophy in the accomplishments of science. It is ironic that these authors are, in fact, endorsing the beliefs of ID proponents who argue that science is inherently materialist, and they share the creationists' conceit that science makes assertions about values, meanings, and purposes.

The National Academy of Sciences has emphatically asserted: "Religion and science answer different questions about the world. Whether there is a purpose to the universe or a purpose for human existence are not questions for science. . . . Consequently, many people including many scientists, hold strong religious beliefs and simultaneously accept the occurrence of evolution."[13]

The Academy also asserts: "Within the Judeo-Christian religions, many people believe that God works through the process of evolution. That is, God has created both a world that is ever-changing and a mechanism through which creatures can adapt to environmental change over time."[14] Among writers who share such belief are biologist and textbook author Kenneth R. Miller, the Catholic theologian John F. Haught, the Episcopalian biochemist and theologian Arthur Peacocke, and so many other theologians and religious authors.[15]

Scientific knowledge cannot contradict religious beliefs, because science has nothing definitive to say for or against religious inspiration, religious realities, or religious values. There are Christian believers, however, who see the theory of evolution and scientific cosmology as contrary to the creation narrative of the book of Genesis. These believers are entitled, of course, to hold such convictions based on their interpretation of Scripture. But Genesis is a book of religious revelations and of religious teachings, not a treatise of astronomy or biology. As I quoted above, Pope John Paul II has stated that the Bible speaks of the origins of the universe and its makeup, "not in order to provide us with a scientific treatise, but in order to state the correct relationships of man with God and the universe."

Augustine and many other religious authorities have made similar statements to the effect that it is a blunder to mistake the Bible for an elementary textbook of astronomy, geology, and other natural sciences. Galileo argued that if reason leads to discovery of a truth that seems to be incompatible with the Bible, then the interpretation of the Bible, not the scientific conclusion, should give way. He was, in fact, following Augustine, who, in his commentary on the book of Genesis, had written: "If it happens that the authority of sacred Scripture is set in opposition to clear and certain reasoning, this must mean that the person who interprets Scripture does not understand it correctly."[16]

It is possible to believe that God created the world while also accepting that the planets, mountains, plants, and animals came about, after the initial creation, by natural processes. In theological parlance, God may act through secondary causes. Similarly, at the personal level of the individual, I can believe that I am God's creature without denying that I developed from a single cell in my mother's womb by natural processes. For the believer the providence of God impacts personal life and world events through natural causes. The point, once again, is that scientific conclusions and religious beliefs concern different sorts of issues, belong to different realms of knowledge; they do not stand in contradiction.

For some Christians, however, the reason to reject evolution and other scientific knowledge is because they maintain that the Bible should be taken literally, in its religious teachings as well as in all historical and descriptive references to the world. These believers, as pointed out earlier, encounter an insurmountable difficulty in the contradictory statements found in the Bible, such as the two inconsistent narratives of the creation of the world and of humankind in chapters 1 and 2 of Genesis.

Science as a Way of Knowing

Science is a wondrously successful way of knowing the world. Science seeks explanations of the natural world by formulating hypotheses that are subject to empirical falsification or corroboration. A scientific hypothesis is tested by ascertaining whether or not predictions about the world of experience derived as logical consequences from the hypothesis agree with what is actually observed.

Science as a mode of inquiry into the nature of the universe has been immensely successful and of great consequence. Witness the proliferation of science academic departments in universities and other research institutions, the enormous budgets that the body politic and the private sector willingly commit to scientific research, and its economic impact. The U.S. Office of Management and Budget has estimated that 50 percent of all economic growth in the United States since the Second World War can be attributed directly to scientific knowledge and technical advances. Indeed, the technology derived from scientific knowledge pervades our lives: the high-rise buildings of our cities, thruways and long-span bridges, rockets that take men and women into outer space, telephones that provide instant communication across continents, computers that perform complex calculations in millionths of a second, vaccines and drugs that keep bacterial parasites at bay, gene therapies that replace DNA in defective cells. All these remarkable achievements bear witness to the validity of the scientific knowledge from which they originated.

Scientific knowledge is also remarkable in the way new knowledge builds upon past accomplishments rather than starting anew with each generation or each new practitioner. Surely scientists disagree with each other on many matters; but these are issues not yet settled, and the points of disagreement generally do not bring into question prior knowledge. Most scientists do not challenge the existence of atoms, or that there is a universe with myriad stars, or that heredity is encased in DNA.

Beyond Science

Science is a way of knowing, but it is not the only way. Knowledge also derives from other sources, such as common sense, artistic and religious experience, and philosophical reflection. In *The Myth of Sisyphus*, the great French writer Albert Camus asserted that we learn more about ourselves and the world from a relaxed evening's perception of the starry heavens and the scents of grass than from science's reductive ways.[17] The anthropologist Loren Eiseley wrote: "The world without Shakespeare's insights is a lesser world, our griefs shut more inarticulately in upon themselves. We grow mute at the thought—just as an element seems to disappear from sunlight without Van Gogh."[18] Astonishing to me is the assertion made by some scientists and others that there is no valid knowledge outside science. I respond with a witticism that I once heard from a friend: "In matters of values, meaning, and purpose, science has all the answers, except the interesting ones."

The validity of the knowledge acquired by nonscientific modes of inquiry can be simply established by pointing out that science (in the modern sense of empirically tested laws and theories) dawned in the sixteenth century, but mankind had for centuries built cities and roads, brought forth political institutions and sophisticated codes of law, advanced profound philosophies and value systems, and created magnificent plastic art, as well as music and literature. The crops we harvest and the animals we husband emerged, millennia before science's dawn, from practices established by farmers in the Middle East, China, Andean sierras, and Mayan plateaus. We learn about the human predicament reading Shakespeare's *King Lear,* contemplating Rembrandt's *Self-Portraits,* and listening to Tchaikovsky's *Symphonie Pathétique* or Elton John's *Candle in the Wind.* We thus learn about ourselves and about the world in which we live, and we also benefit from products of this nonscientific knowledge. We humans have systems of morality concerning the consequences of our actions in regard to others, and derive meaning and purpose from religious beliefs.

It is not my intention here to belabor the extraordinary fruits of nonscientific modes of inquiry. I wish simply to state something that is obvious, but at times becomes clouded by the hubris of some scientists. Successful as it is, and universally encompassing as its subject is, a scientific view of the world is hopelessly incomplete. Matters of value and meaning are outside science's scope. Even when we have a satisfying scientific understanding of a natural object or process, we are still missing matters that may well be thought by many to be of equal or greater import. Scientific knowledge may enrich esthetic and moral perceptions and illuminate the significance of life and the world, but these concerns are outside science's realm.

I reiterate my convictions by quoting the words of Freeman Dyson, a distinguished scientist and writer whom I much admire: "As human beings, we are groping for knowledge and understanding of the strange universe into which we are born. We have many ways of understanding, of which science is only one. . . . Science is a particular bunch of tools that have been conspicuously successful for understanding and manipulating the material universe. Religion is another bunch of tools, giving us hints of a mental or spiritual universe that transcends the material universe."[19]

In the final chapter of this book, I consider historical and philosophical issues that may not be of interest to all readers. I list the essential features of the scientific method and how these were misinterpreted by some influential philosophers in Darwin's time and later, which explains why Darwin was credited with the theory of evolution, rather than with the theory of natural selection.

10

POSTSCRIPT FOR THE COGNOSCENTI

> [I]t occurred to me, in 1837, that something might per-
> haps be made out on this question [the origin of species]
> by patiently accumulating and reflecting on all sorts of facts
> which could possibly have any bearing on it.
>
> Charles Darwin, *The Origin of Species*, p. 1

I assert in Chapter 3 that Darwin's most significant contribution to science is not the demonstration of the evolution of
organisms, but his discovery of natural selection—a process
that explains the "design" of organisms and their harmonious
complexity, in addition to their change and diversification
through eons of time. I further aver that Darwin's discovery of
natural selection is one of the most significant achievements in
intellectual history because it completed the Copernican Revolution by bringing the magnificent adaptations of organisms
and their amazing diversity to the realm of science: explanation
by natural laws and processes.

The Copernican Revolution represents the first major development in the history of science; the Darwinian Revolution is
the second and definitive one. Copernicus, Galileo, and Newton
brought the world of inanimate nature, on Earth as well as in
the heavens, to the domain of science, thereby excluding super-

natural explanations. Darwin brought the world of life, with all its diversity and splendid contrivances, to the domain of science, making unnecessary the awkward recourse to a designer who would again and again intervene in the natural world with designs that are often imperfect and occasionally dysfunctional.

Darwin's theory, by Darwin's own assessment, was the theory of natural selection, much more so than the theory of evolution. Why did history shift the emphasis from natural selection, the causative *process,* to evolution, the *events* or outcomes of the process? The explanation comes from the historical context within which Darwin presented his discoveries; how science and the scientific process were understood in Darwin's time and the ensuing decades. In the mid-nineteenth century, particularly but not only in Great Britain, the prevailing philosophy was empiricism. Empiricism proclaimed that science proceeds by "induction" from observed facts to universal laws of ever greater generality. Comprehensive science is to be approached step by step, with conclusions and laws always emerging from observed facts, but avoiding rational, or speculative, constructs.

In this final chapter, I first delineate the process of induction as understood in Darwin's time. I then argue that the scientific method, as practiced by scientists, is not essentially a process of induction, but consists rather of two episodes. The first episode consists primarily of the formulation of a hypothesis; the second episode consists of experimentally testing the hypothesis. Induction often plays a role in the formulation of scientific hypotheses, as well as in the acquisition of other sorts of knowledge, including common sense, aesthetics, and philosophy.

What differentiates science from other forms of knowledge is the second episode: subjecting hypotheses to empirical testing, to falsification or corroboration by observing whether or not predictions derived from a hypothesis are the case in relevant observations and experiments.

After a brief discussion of the scientific method, I review Darwin's scientific practice, how he proceeded in his research. I point out that there is a flagrant contradiction between Darwin's methodology and how he described it for public consumption. Darwin's experimental investigations and the books in which he reported them are relentless efforts to test the hypothesis of natural selection, precisely by studying cases in which it seemed most unlikely to obtain, as in orchids, barnacles, and earthworms. For the public, however, he claimed to follow the canons of induction.

Last, I briefly examine the scientific method as practiced by Gregor Mendel, a younger contemporary of Darwin who discovered the theory of biological heredity. Mendel's scientific practice was also directed toward developing a hypothesis, a theory of inheritance and, then, testing it. Yet his discoveries became known as the "laws" of heredity, rather than as the "theory" of heredity. As in the case of Darwin, the temper of the times saw Mendel's discoveries as induction of general laws from observed facts, rather than as the construction of a powerful theory, which Mendel's theory was. Mendel devised clever experiments to test his theory. For decades thereafter, Mendel's theory of heredity (with some modifications and many extensions) inspired numerous developments and has remained the core of genetics, surely one of the most successful scientific disciplines of the twentieth century and into the present. Mendel's theory of heredity is particularly relevant to the message of this book because it provided the "missing link" in Darwin's own theory of evolution by natural selection.

Induction and Empiricism

It is a common misconception that science advances by accumulating experimental facts and drawing up general laws from them. This misconception is encased in the much-repeated assertion that science is inductive, a notion that can be traced to the English statesman and essayist Francis Bacon (1561–1626). Bacon had an important and influential role in shaping modern science by his criticism of the prevailing metaphysical speculations of medieval scholastic philosophers. In the nineteenth century, the most articulate and influential proponent of inductivism was John Stuart Mill (1806–1873), an English philosopher and economist.

Induction was proposed by Bacon and Mill as a method of achieving "objectivity," while avoiding subjective preconceptions, and of obtaining "empirical" rather than abstract or metaphysical knowledge. In its extreme form, this proposal would hold that a scientist should observe any phenomena that he encounters in his experience and record the observations without any preconception as to what the truth about them might be. Truths of universal validity are expected to emerge eventually, as a result of the relentless accumulation of unprejudged observations. The methodology proposed may be exemplified as follows. A scientist measuring and recording everything that confronts him observes a tree with leaves. A second tree, and a third, and many others, are all observed to have leaves. Eventually, he formulates a universal statement, "All trees have leaves."[1]

The inductive method succeeded, according to Bacon, because knowledge ascended by the "ladder of the intellect," from minor but careful observations to general conclusions that had to be true because they were grounded on the direct experiences of nature. Mill did not recognize deductive reasoning as a source of knowledge. Rather, he believed that all valid propositions are either reports on experience or generalizations from experience. He even claimed that the propositions of mathematics are merely very large scale empirical generalizations. Logical deduction does not yield new knowledge. Rational inference amounts simply to explicative verbal inference.

But Mill was mistaken: scientific theories are not established by induction. The inductive method fails to account for the actual process of science. First, no scientist works without any preconceived plan as to what kind of phenomena to observe. Scientists choose for study objects or events that, in their opinion, are likely to provide answers to questions that interest them. Otherwise, as Darwin wrote, "A man might as well go into a gravel pit and count the pebbles and describe the colours."[2] A scientist whose goal was to carefully record events observed in all waking moments of his life, without focusing on those that would interest him, would not contribute much to the advance of science; more likely than not, he would be considered mad by his colleagues.

Moreover, induction fails to arrive at universal truths. No matter how many singular observations may be accumulated, no universal statement can be logically derived from such an accumulation. Even if all trees so far observed have leaves, or all swans observed are white, it remains a logical possibility that the next tree will not have leaves, or the next swan will not be white. The step from numerous singular statements to a universal one involves logical amplification. The universal statement has greater logical content—it says more—than the sum of all singular statements.

Another serious logical difficulty with the proposal that induction is "the" method of science, is that scientific

hypotheses and theories are formulated in abstract terms
that do not occur at all in the description of empirical events.
Mendel, the founder of genetics, observed in the progeny of
hybrid plants that alternative traits segregated according to certain proportions. Repeated observations of these proportions
could never have led inductively to "observe" the "factors" (now
called "genes") formulated in his hypothesis or their presence
and segregation in the sex cells. The genes were not observed
and thus could not be included in statements reflecting what
Mendel observed. The most interesting and fruitful scientific
hypotheses are not simple generalizations. Instead, scientific
hypotheses are creations of the mind, imaginative suggestions
as to what might be true.

Induction fails on all three counts pointed out. It is not
a method that ensures objectivity and avoids preconceptions,
it is not a method to reach universal truths, and it is not a
good description of the process by which scientists formulate
hypotheses and other forms of scientific knowledge. It is a different matter that a scientist may come upon a new idea or
develop a hypothesis as a consequence of repeated observations
of phenomena that might be similar or share certain traits. But
how we come upon a new idea is quite a different matter from
how it is that we come to accept an idea as established scientific
knowledge. I come back to this point later.

Science as Knowledge

Three characteristic traits jointly distinguish scientific knowledge from other forms of knowledge.[3] First, science seeks
the *systematic* organization of knowledge about the world.
Common sense provides knowledge about natural phenomena,
and this knowledge is often correct. For example, common

sense tells one that children resemble their parents and that good seeds produce good crops. Common sense, however, shows little interest in systematically establishing connections between phenomena that do not appear to be obviously related. Figurative art and imaginative literature are valid explorations of knowledge about the world, but art and literature, like common sense, do not seek systematic explanations of the realities they explore. By contrast, science is concerned with formulating general laws and theories that manifest patterns of relations between very different kinds of phenomena. Science develops by discovering new relationships, and particularly by integrating statements, laws, and theories, which previously seemed to be unrelated, into more comprehensive laws and theories.

Second, science strives to explain *why* observed events do in fact occur. The knowledge acquired in the course of ordinary or aesthetic experience is frequently accurate, but it seldom provides explanations of why phenomena occur as they do. Practical experience tells us that children resemble one parent in some traits and the other parent in other traits, or that manure increases crop yield. Poetry, music, and representational art provide profound and meaningful insights into human nature and the significance of life and the world. But neither common knowledge nor aesthetic experience provides explanatory accounts of knowledge. Science, on the other hand, seeks to formulate explanations for natural phenomena by identifying the conditions that account for their occurrence.

Seeking the systematic organization of knowledge and trying to explain why events are as observed are two characteristics that distinguish science from commonsense knowledge and from aesthetic experience. These characteristics are also shared by other forms of systematic knowledge, such as mathematics, philosophy, and theology. A third characteristic of science, and the one that distinguishes the empirical sciences from other systematic forms of knowledge, is that scientific explanations must be formulated in such a way that they can be subjected to empirical *testing*, a process that must include the possibility of "empirical falsification," a notion I explain later.

New ideas in science are advanced in the form of hypotheses. Hypotheses are mental constructs, imaginative exploits, that provide guidance as to what is worth observing and that encourage the scientist to seek observations that would corroborate or falsify the hypothesis. The tests to which scientific ideas are subjected include contrasting hypotheses with the world of experience in a manner that must leave open the possibility that anyone might reject any particular hypothesis if it leads to wrong predictions about the world of experience. The possibility of empirical falsification of a hypothesis is carried out by ascertaining whether or not precise predictions derived as logical consequences from the hypothesis agree with the state of affairs found in the empirical world. A hypothesis that cannot be subjected to the possibility of rejection by observation and experiment cannot be regarded as scientific. The possibility of empirical falsification of its hypotheses has been called by the philosopher Karl Popper the "criterion of demarcation" that sets scientific knowledge apart from other forms of knowledge.[4]

Science's Validation of Knowledge

Science is a complex enterprise that essentially consists of two interdependent episodes, one imaginative or creative, the other critical. To have an idea, advance a hypothesis, or suggest what might be true is a creative exercise, but scientific conjectures or hypotheses must also be subject to critical examination and empirical testing. Scientific thinking may be characterized as a process of invention or discovery followed by validation or confirmation. One episode concerns the formulation of new ideas (sometimes referred to as the "acquisition of knowledge"); the other concerns their validation ("justification of knowledge").[5]

Scientists, like other people, come upon new ideas and acquire knowledge in all sorts of ways: from conversation with other people, from reading books and newspapers, from inductive generalizations, and even from dreams and mistaken observations. Newton is said to have been inspired by a falling apple. The German chemist August Kekulé (1829–1896) had been unsuccessfully attempting to devise a model for the molecular structure of benzene. One evening he was dozing in front of the fire. The flames appeared to Kekulé as snake-like arrays of atoms. Suddenly one snake appeared to bite its own tail and then whirled mockingly in front of him. The circular appearance of the image inspired in him the model of benzene as a hexagonal ring. Natural selection as the explanation of design came to Darwin while riding in his coach and observing the countryside: "I can remember the very spot in the road . . . when to my joy the solution came to me. . . . The solution, as I believe, is that the modified offspring . . . tend to become adapted to many and highly diversified places in the economy of nature."[6]

Hypotheses and other imaginative exploits are the initial stage of scientific inquiry. It is the imaginative conjecture of what might be true that provides the incentive to seek the truth and a clue as to where we might find it.[7] Hypotheses guide observation and experiment because they suggest what to observe. The empirical work of scientists is guided by hypotheses, whether explicitly formulated or simply in the form of vague conjectures or hunches about what the truth might be. But imaginative conjecture and empirical observation are mutually interdependent episodes. Observations made to test a hypothesis are often the inspiring source of new conjectures or hypotheses. The results of an experiment often inspire the modification of a hypothesis and the design of new experiments to test it.[8]

The starting point of scientific inquiry is the conception of an idea. The creative process is, however, not unique to scientists. Philosophers, as well as novelists, poets, and painters, are also creative; they too advance models of experience and they also generalize by induction. What distinguishes science from other forms of knowledge is the process by which this knowledge is justified or validated.

The validity of a hypothesis depends on several factors, such as whether it has explanatory value (i.e., makes observed phenomena intelligible in some sense), and whether it is consistent with hypotheses and theories commonly accepted in the particular field of science, although some of the greatest scientific advances occur precisely when it is shown that a widely held hypothesis is replaced by a new one that accounts for the phenomena explained by the preexisting hypothesis, as well as additional phenomena.[9]

The most distinctive feature of the scientific process is the need to put on trial every scientific hypothesis by ascertaining whether or not predictions about the world of experience derived from the hypothesis agree with what is actually observed. The requirement that scientific hypotheses be empirically falsifiable is the critical element that distinguishes the empirical sciences from other forms of knowledge. Scientific hypotheses cannot be consistent with all possible states of affairs in the empirical world. A hypothesis is scientific only if it is consistent with some but not with other possible states of affairs not yet observed in the world, so that it may be subject to the possibility of falsification by observation. The predictions derived from a scientific hypothesis must be sufficiently precise that they limit the range of possible observations with which they are compatible. If the results of an empirical test agree with the predictions derived from a hypothesis, the hypothesis is said to be provisionally corroborated; otherwise it is falsified. A hypothesis that is not subject to the possibility of empirical falsification does not belong in the realm of science.[10]

Scientists gain confidence in the validity of a hypothesis when strenuous efforts—observations and experiments—to falsify the hypothesis fail. The confidence in a particular hypothesis further increases when other hypotheses are built upon the foundations of the former hypothesis, and these new hypotheses also stand the test of observation and experimentation.

Darwin and Empiricism

The model of scientific practice that I have sketched is evident in the practice of eminent scientists who experimentally tested predictions derived from their scientific theories. Examples are, in the seventeenth century, Galileo's (1564–1642) laws of motion, William Harvey's (1578–1657) theory of the circulation of the blood, Blaise Pascal's (1623–1662) explanation of atmospheric pressure, and Isaac Newton's (1642–1727) theories of mechanics and optics; in the eighteenth century Antoine-Laurent Lavoisier's (1743–1794) theory of oxidation; and among nineteenth-century biologists, Claude Bernard (1813–1878), Louis Pasteur (1822–1895) and Gregor Mendel (1822–1884), as well as Darwin himself, as we shall see. There was often, however, a disconnect, particularly in English-speaking countries, between what scientists did and what they pretended to be doing, which they asserted in order to satisfy the expectations of preeminent contemporary philosophers of empiricist persuasion. Notable is the case of Newton, the great-est theorist among scientists, who proclaimed *hypotheses non fingo* ("I fabricate no hypotheses").

In his *Autobiography,* Darwin asserts that he proceeded "on true Baconian principles and without any theory collected facts on a wholesale scale."[11] The opening paragraph of Darwin's *The Origin of Species,* partially quoted at the beginning of this chapter, conveys the same impression. Let me take that citation from the beginning of the paragraph:

> When on Board H.M.S. *Beagle,* as naturalist, I was much struck with certain facts in the distribution of the inhabit-ants of South America, and in the geological relations of the present to the past inhabitants of that continent. These facts seemed to me to throw some light on the origin of species— that mystery of mysteries, as it has been called by one of our greatest philosophers. On my return home, it occurred to me, in 1837, that something might perhaps be made out on this question by *patiently accumulating and reflecting on all sorts of facts which could possibly have any bearing on it.* After five years' work I allowed myself to speculate on the subject, and drew up some short notes; these I enlarged in 1844 into a sketch of the conclusions, which then seemed to me prob-able: from that period to the present day I have steadily pur-sued the same object.[12]

Darwin claimed also in other writings to have followed the inductivist canon.

The facts are very different from these claims, however. Darwin's notebooks and private correspondence show that he came upon the hypothesis of natural selection in 1837, several years before he claims to have allowed himself for the first time "to speculate on the subject." Between shortly after the return of the *Beagle* on October 2, 1836, and the publication of *The Origin* in 1859 (and, indeed, until the end of his life), Darwin relentlessly pursued empirical evidence to test his theory of natural selection and to corroborate the evolutionary origin of organisms in support of his theory.

Why this disparity between what Darwin was doing and what he claimed? There are at least two reasons. First, in the temper of the times, "hypothesis" was a term often reserved for metaphysical speculations without empirical substance, as shown by Newton's dictum cited above. Darwin also expressed distaste and even contempt for empirically untestable hypotheses. He wrote of Herbert Spencer: "His deductive manner of treating any subject is wholly opposed to my frame of mind. His conclusions never convince me. . . . His fundamental generalizations (which have been compared in importance by some persons with Newton's Laws!) which I daresay may be very valuable under a philosophical point of view, are of such a nature that they do not seem to me to be of any strictly scientific use. They partake more of the nature of definitions than of laws of nature. They do not aid me in *predicting what will happen* in any particular case."[13]

There is another reason, a tactical one, why Darwin claimed to proceed according to inductive canons: he did not want to be accused of subjective bias in the evaluation of empirical evidence. Darwin's true colors are shown in a letter to a young scientist written in 1863: "I would suggest to you the advantage, at present, of being very sparing in introducing theory in your papers (I formerly erred much in Geology in that way): *let theory guide your observations*, but till your reputation is well established, be sparing of publishing theory. It makes persons doubt your observations."[14]

Darwin rejected the inductivist claim that observations should not be guided by hypotheses. The statement I quoted earlier, "A man might as well go into a gravel-pit and count the pebbles and describe the colours," is followed by this tell-ing remark: "How odd it is that anyone should not see that all observation must be for or against some view if it is to be of any service!"[15] He acknowledged the heuristic role of hypotheses, which guide empirical research by telling us what is worth observing, what evidence to seek. He confesses: "I cannot avoid forming one [hypothesis] on every subject."[16]

Modern students of Darwin have abundantly shown that Darwin was an excellent practitioner of the hypothetico-deductive method of science.[17] Darwin advanced hypotheses in multiple fields, including geology, plant morphology and physiology, psychology, and evolution, and subjected his hypotheses to empirical tests. "The line of argument often pur-sued throughout my theory is to establish a point as a prob-ability by induction and to apply it as a hypothesis to other parts and see whether it will solve them."[18] Darwin recognized that falsification of seemingly true hypotheses contributes to the advancement of science: "False facts are highly injurious to the progress of science, for they often endure long; but false views, if supported by some evidence, do little harm, for every one takes a salutary pleasure in proving their falseness; and when this is done, one path towards error is closed and the road to truth is often at the same time opened."[19]

There can be little doubt that natural selection and other causal processes of evolution are investigated by formulation and empirical testing of hypotheses. The study of evolution-ary history is also based on the formulation of empirically testable hypotheses. Consider a simple example. For many years specialists proposed that the evolutionary lineage lead-ing to man separated from the lineage leading to the great apes (chimpanzee, gorilla, orangutan) before the lineages of the great apes separated from each other. Some recent authors have suggested instead that man and chimpanzee are more closely related to each other than the chimpanzee is to the gorilla or the orangutan. A wealth of empirical predictions can be derived logically from these competing hypotheses. One prediction concerns the degree of similarity between the DNA and the proteins of these species. It is known that DNA and proteins accumulate gradually over time. If the older hypothesis is correct, the average amount of DNA and protein differentiation should be greater between man and the great apes than among these. On the other hand, if the newer hypothesis is correct, man and chimpanzee should exhibit greater similarity than either one with the other apes. These alternative predictions provide a critical empirical test of the hypotheses. The available data favor the second hypothesis. Man and chimpanzee appear to be more closely related to each other than they are to the gorilla or to the orangutan.

A Historical Paradigm: Mendel's Theory of Genetics

Before I finish this chapter, I want to consider the case of Gregor Mendel, who formulated the theory upon which modern genetics is built. Mendel's story is relevant here for two reasons: First, it illustrates the same situation as that of Darwin's, in that his main achievement was interpreted to be the formulation of laws, obtained by inductive generalization, rather than his theory of biological heredity. Second, Mendel, as noted earlier in the book, provided the missing link in Darwin's theory of natural selection. This allows me to speculate, at the very end of this chapter, that if Mendel's theory had been known in Darwin's time (it was published in 1866, 7 years after the publication of *The Origin,* but it did not become generally known to scientists until the start of the twentieth century), Darwin's theory of natural selection would have been readily accepted by scientists and perhaps, thereby, the antagonism between creationists and evolutionists might have been, at least in good part, avoided.

Mendel conducted experiments that yielded regularities, or "laws," in the transmission of traits from parents to offspring. On the basis of these observations, he formulated a theory of heredity, which he then proceeded to test with clever, critical experiments. Mendel's discoveries became known in the twentieth century primarily not as a theory of heredity but as the "laws" of heredity. The prevailing empiricism and its pernicious influence in the practice and teaching of scientists was the cause of this distortion, which persists in current textbooks of genetics. (I hasten to add that teaching "Mendel's laws" has not kept geneticists from practicing their science by advancing hypotheses and testing them experimentally.)

Gregor Mendel was an Augustinian monk living in the Austrian city of Brünn (now Brno, Czech Republic). Mendel tackled the problem of biological heredity and succeeded where better known contemporary scientists and distinguished predecessors had failed. He performed experiments with pea plants and reported his discoveries in a paper published in 1866, "Experiments in Plant Hybridization," remarkable for his lucid awareness of the requirements of the scientific method.[20] Mendel formulated hypotheses, examined their consistency with previous results, and then submitted the hypotheses to severe critical tests and suggested additional tests that might be performed.

Mendel's genius is evident in his recognition of the conditions required to formulate and test a theory of inheritance: different traits in a plant (such as flower color or seed shape) should be considered individually, alternative states of the traits should differ in clear-cut ways (such as white and purple flower color), and ancestry of the plants should be precisely known by using only true breeding lines in the experiments. Mendel's preliminary inferences (his "laws") were formulated in probabilistic terms; accordingly, he obtained large samples and subjected them to statistical analysis.

Mendel studied the transmission of seven different traits in the garden pea (*Pisum sativum*) including the color of the seed (yellow versus green) and of the flower (white versus purple), the configuration of the seed (round versus wrinkled), and the height of the plant (tall versus dwarf). Mendel's first series of experiments was with plants that differ in a single trait. The results led him to generalizations formulated as "laws": Only one of the two traits (the "dominant" trait, yellow seed, for example) appears in the first-generation progenies. After self-fertilization, three-fourths of the second-generation progenies exhibit the dominant trait, and one-fourth exhibit the other ("recessive" trait, green seed, for example). The second-generation plants exhibiting the recessive trait breed true in the following generations (i.e., all green-seed plants produce green-seed plants), but the plants exhibiting the dominant trait are of two kinds—one-third breed true (i.e., one-third yellow-seed plants produce only yellow-seed plants), the other two-thirds are hybrids (i.e., produce both yellow-seed and green-seed plants). Mendel tested these observations by repeating his experiments for each of the seven characters. The ensuing generalization was summarized in a principle, later called the law of segregation: hybrid plants produce seeds that are one-half hybrid, one-fourth pure breeding for the dominant trait, and one-fourth pure breeding for the recessive trait.

The study of the offspring of crosses between plants differing in two traits (e.g., round and yellow seeds in one parent, wrinkled and green seeds in the other parent) led him to for-

mulate a second generalization, later called the law of independent assortment: "The principle applies that in the offspring of
the hybrids in which several essentially different characters are
combined, . . . the relation of each pair of different characters
in hybrid union is independent of the other differences in the
two original parental stocks." He corroborated this principle by
examining progenies of plants differing in three and four traits.
He correctly predicted and corroborated experimentally that in
the progenies of plants hybrid for n characters there will be 3^n
different classes of plants.

The formulation and experimental testing of the two principles take up only the first half of Mendel's paper. Midway
through the paper, Mendel advances what he properly calls a
"hypothesis" or theory to account for his previous results and
for the two laws. The second half of the paper is dedicated to the
derivation of predictions from the theory and to testing them.

Mendel's theory of inheritance contains the following elements: (1) for each character in any plant, whether hybrid or
not, there is a pair of hereditary factors ("genes"); (2) these two
factors are inherited, one from each parent; (3) the two factors
of each pair segregate during the formation of the sex cells,
so that each sex cell receives only one factor; (4) each sex cell
receives one or the other factor of a pair with a probability of
one-half; (5) alternative factors for different characters associ-
ate at random in the formation of the sex cells.

Mendel's well-deserved eminence as one of the great scientists of all times rests particularly on the formulation of
this theory of heredity. Mendel was also quite aware of the
epistemological status of his proposal, namely that it was a
hypothesis that required experimental corroboration. Just
after formulating the theory that I have summarized in the
preceding paragraph, Mendel wrote that "this hypothesis

would fully suffice to account for the development of the hybrids in the separate generations," that is, the hypothesis is consistent with his previous experiments. But that was not enough, as he recognized, since the theory had been designed to fit the results. New tests would be required. He writes: "In order to bring these assumptions to an experimental proof the following experiments were designed." The tests are two series of ingenious experiments, particularly the so-called "back-crosses" that predict, if his theory is correct, a distribution of characters among the progeny in proportions radically different from those he had previously observed. He conducted these experiments, and the results corroborated the key elements of his theory.

Yet, Mendel passed to history, courtesy of empiricism, as the discoverer of the two "laws" of inheritance, the "law of segregation" and the "law of independent assortment," validated by repeated observations, rather than as the proponent of a theory of biological inheritance, which he suitably corroborated and which has, by and large, remained the core of the modern theory of genetics.

Epilogue: A Historical Footnote

History abounds in footnotes that show how seemingly trivial events could have changed the course of history if they would have happened differently. One event that involves Darwin and Mendel is apposite here. I pointed out in Chapter 4 that the "missing link" in Darwin's theory of natural selection was the lack of an adequate theory of heredity. Theories of inheritance prevailing in the nineteenth century, including Darwin's own "provisional hypothesis of pangenesis,"[21] were inconsistent with his theory of natural selection. Yet, in Darwin's time, less than a decade after the publication of *The Origin*, Gregor Mendel had formulated a theory of inheritance, which maintains currency in its fundamentals to the present, that was consistent with Darwin's explanation of biological design and evolution by natural selection.

Mendel first learned of Darwin's theory most likely in September 1861, when K. Schwippel lectured in Brno "On the geological succession of organic beings."[22] A German translation of *The Origin* was acquired in 1862 by the Natural Science Society of Brno, where Mendel was an assiduous participant in the monthly meetings and where three years later, in 1865, he would present his "Experiments in Plant Hybridization," published the following year in the Bulletin of the Society. A copy of the second German edition of *The Origin*, published in 1863 was bought by Mendel and contains his marginal notations. There are also marginal annotations by Mendel in his copy of the German translation, published in 1868, of Darwin's *Variation in Animals and Plants under Domestication*. The marginalia are quite extensive in the second volume, where Darwin develops his pangenesis hypothesis.[23] Mendel repeatedly points out serious flaws in Darwin's proposed explanation that each adult organ produces tiny particles, which Darwin called "gemmules," representative of that organ's particular features. According to Darwin's theory, particles from different organs collect together in the sex cells and are passed on to offspring. A fatal flaw of this theory with respect to natural selection is that any difference between the parents would be halved each generation, so that the natural selection advantage of a new variant would be reduced generation after generation and become virtually obliterated.

There is every reason to assume that Darwin would have understood the significance of Mendel's theory of heredity, which involved segregation of the parental traits in the hybrids, since it would have solved this important difficulty in Darwin's theory. [24] It seems, or so the story has been repeatedly told, that Mendel had sent to Darwin a reprint of his hybridization paper, which, however, was found unopened in Darwin's library after his death. This is hardly surprising. Although Mendel had correspondence with some of the greatest contemporary botanists on the Continent, notably Carl Nägeli (1817–1891), he was all but unknown to British scientists and surely to Darwin.

The history of biology might have been significantly different had Darwin read Mendel's paper. Darwin's theory of natural selection was widely challenged, or even ignored, in the late 1800s and the early decades of the twentieth century, largely because of its inconsistency with current ideas about biological inheritance. [25] Eventually, after the rediscovery of Mendel's work, it was shown that natural selection was compatible with biological heredity. One may surmise that had Darwin incorporated Mendel's theory into his own, natural selection might have become generally accepted much earlier than it was. Whether or not an early acceptance of Darwin's theory by the scientific community at large might have forestalled the arising antagonism between evolution and religion, we cannot tell.

Be that as it may, this speculative possibility might be a good point at which to end my own efforts to persuade people of faith as well as other readers that there need be no antagonism between evolution and religious beliefs. People of faith may see the presence of God in the operations of nature and the creative powers of natural selection, first discerned by Darwin. "There is grandeur in this view of life." [26]

The theory of evolution conveys chance and necessity jointly enmeshed in the stuff of life; randomness and determinism interlocked in a natural process that has spurted the most complex, diverse, and beautiful entities in the universe: the organisms that populate the Earth, including humans who think and love, endowed with free will and creative powers, and able to analyze the process of evolution itself that brought them into existence. This is Darwin's fundamental discovery, that there is a process that is creative though not conscious. And this is the conceptual revolution that Darwin completed: the idea that the design of living organisms can be accounted for as the result of natural processes governed by natural laws. This is nothing if not a fundamental vision that has forever changed how mankind perceives itself and its place in the universe.

NOTES

CHAPTER 1

1. Thomas Aquinas, *Summa Theologiae*, I, 2, 3. A bilingual edition in English and Latin, in 60 volumes, has been published by Blackfriars & McGraw-Hill. *Existence and Nature of God* I, 2–11 (1964) is vol. 2. In the *Summa Contra Gentiles*, Aquinas advances the same argument, which derives, he says, from St. John Damascene: "It is impossible for contrary and discordant things to fall into one harmonious order except under some one guidance, assigning to each and all parts a tendency to a fixed end. But in the world we see things of different natures falling into harmonious order.... Therefore there must be some Power by whose providence the world is governed; and that we call God" ("God and His Creatures, Book I").

2. The Church of Christ, Scientist's solution to the dilemma is to deny the existence of evil, which is said to be nothing but an illusion. This solution is contrary to the Christian tradition. Already

before Augustine's (354–430) formulation of original sin, the Latin Fathers of the Christian Church and, even earlier, Irenaeus (130–200) and other Greek-speaking Fathers had accepted the Fall and the reality of sin.

Chapter 2

1. William Paley, *Natural Theology* (New York: American Tract Society, n.d.), 15–16. I cite pages following this American edition, which is undated but seems to have been printed in the late nineteenth century.
2. Ibid., 20.
3. Ibid.
4. Ibid., 22–23.
5. Ibid., 23.
6. Ibid., 48.
7. Ibid.
8. Ibid.
9. Ibid., 1.
10. See M. Behe, *Darwin's Black Box: The Biochemical Challenge to Evolution* (New York: The Free Press, 1996), Chapter 7.
11. Paley, *Natural Theology*, 175–176.
12. Ibid., 180, 183.
13. Ibid., 265.
14. Ibid., 47.
15. Ibid.
16. Ibid.
17. Ibid., 46.
18. Augustine, *The City of God*, edited and translated by R. W. Dyson (Cambridge, UK: Cambridge University Press, 1998), 452–453. As quoted by M. Ruse, "The Argument from Design," in W. A. Dembski and M. Ruse, eds., *Debating Design: From Darwin to DNA* (Cambridge, UK: Cambridge University Press, 2004), 13–31.
19. T. Aquinas, *Summa Theologiae*, Part I, 2, 3; vol. 2, 27 (see note 1 of Chapter 1).

20. J. Ray, *The Wisdom of God, Manifested in the Works of Creation* (London, 1691), 33.

21. See "Voltaire, François-Marie Arouet de," in *The Encyclopedia of Philosophy*, vol. 8 (London: Macmillan, 1967), 262–270.

22. See M. Roberts, "Intelligent Design. Some Geological, Historical, and Theological Questions" in *Debating Design: From Darwin to DNA*, eds. W. A. Dembski and M. Ruse (Cambridge, UK: Cambridge University Press, 2004), 282.

23. As quoted by G. G. Simpson, *This View of Life* (New York: Harcourt, Brace and World, 1964), 192.

24. F. Darwin (ed.), *Charles Darwin's Autobiography* (New York: Schuman, 1961), 34–35.

CHAPTER 3

1. *On the Origin of Species. A Facsimile of the First Edition, with an Introduction by Ernst Mayr* (New York: Athenaeum, 1967), 489–490 (emphasis added).

2. C. Darwin, *The Autobiography of Charles Darwin (1809–1882)*, ed. N. Barlow (London: Collins, 1958).

3. Sigmund Freud (1856–1939) refers to these two revolutions as "outrages" inflicted upon humankind's self-image:

   Humanity in the course of time had to endure from the hands of science two great outrages upon its naïve self-love. The first was when it realized that our earth was not the centre of the universe, but only a tiny speck in a world-system of a magnitude hardly conceivable; this is associated in our minds with the name of Copernicus, although Alexandrian doctrines taught something very similar. The second was when biological research robbed man of his peculiar privilege of having been specially created, and relegated him to a descent from the animal world, implying an ineradicable animal nature in him: this transvaluation has been accomplished in our own time upon the instigation of Charles Darwin, Wallace, and their predecessors, and not without the most violent opposition from their contemporaries.

S. A Freud, *A General Introduction to Psycho-Analysis* (first published in English in 1920), in *Great Books of the Western World*, vol. 54, *Freud*, ed. M. J. Adler (Chicago: Encyclopædia Britannica, Inc., 1993), 562. Freud proceeds to assert that "the third and most bitter blow" upon "man's craving for grandiosity" was meted out in the twentieth century by psychoanalysis, revealing that man's *ego* "is not even master in his own house." Ibid.

4. Laws such as $f = m \times a$, *force* = *mass* × *acceleration*; or the inverse-square law of attraction, $f = g(m_1 m_2)/r^2$, the force of attraction between two bodies is directly proportional to their masses, but inversely related to the square of the distance between them.

5. For a review of the Red and Transmutation B to E Notebooks, see Chapter 3 in N. Eldredge, *Darwin* (New York: Norton, 2005), 71–138.

6. Wallace's essay was published in the *Journal of the Proceedings of the Linnean Society of London (Zoology)* 3 (1858): 53–62. Darwin's two papers were also published in the same issue.

7. In his paper, Wallace writes: "We believe that there is a tendency in nature to the continued progression of certain classes of varieties further and further from the original type—a progression to which there appears no reason to assign any definite limits. This progression, by minute steps, in various directions. . . ." Ibid., 53.

8. Darwin, *The Origin of Species*, 306. The Silurian period spans from 416 to 444 million years ago.

CHAPTER 4

1. Darwin, *The Origin of Species*, 80–81.

2. For simplicity, I use the term "mutation" to encompass not only changes within a single gene, but also inversions, translocations, and other reorganizations of the genetic material, as well as duplications or deletions, whether of single genes or of gene complexes. Similarly, when I refer to genetic variants, I encompass not only variations of a single gene, but also variations of gene arrangements and gene complexes. Genetic textbooks elucidate these complexi-

ties, and my simplifications do not adversely impact the points made in this book.

3. It is remarkable that the Mendelian theory of heredity, which would become generally known only several decades after publication of *The Origin*, provided a satisfactory account of evolution by natural selection, which could not be achieved by the theories of inheritance prevailing in Darwin's time, including his own pangenesis hypothesis, developed in the second volume of *The Variation of Animals and Plants under Domestication* (London: Murray, 1868, 411, 486). One of the most precious gifts I have ever received is a mint-condition first edition of this work.

4. The range of mutation rates is actually very broad, from more than one in a hundred thousand to less than one in a hundred million. Moreover, there are different ways of measuring mutation rates: for example, rates with respect to changes in any given *letter* of the DNA sequence of a gene, or with respect to any change in any given *gene* (which encompasses hundreds or thousands of DNA letters). Also, rates are quite different for gene mutations in the strict sense and for reorganizations, duplications, and deletions of sets of genes. Genetic textbooks provide more detailed information.

5. M. W. Nachman, H. E. Hoekstra, and S. L. D'Agostino, "The Genetic Basis of Adaptive Melanism in Pocket Mice," *Proceedings of the National Academy of Sciences, USA* 100 (2003): 5268–5273.

6. D. N. Reznick, M. Mateos, and M. S. Springer, "Independent Origins and Rapid Evolution of the Placenta in the Fish Genus *Poeciliopsis*," *Science* 298 (2002): 1018–1020.

7. A useful review of the speciation processes can be found in the *Encyclopaedia Britannica* (Vol. 18, 2005) article on "Evolution, The Theory of," particularly pages 876–886.

8. The evolution from unicellular to multicellular organisms is well understood in the case, for example, of *Volvox* and related species called Volvocales, in which only three or four genes account for the formation of the multicellular aggregate and the differentiation between somatic and reproductive cells.

9. At high levels of the SF-1 protein, enzymes are produced that synthesize testosterone, and males result. At lower levels of the SF-1

protein, an enzyme is made that converts testosterone to estrogen, and females result.

10. An extensive review of the Baldwin effect and other evolutionary processes yielding "novelties" is M. J. West-Eberhard, *Developmental Plasticity and Evolution* (Oxford and New York: Oxford University Press, 2003). See also M. W. Kirschner and J. C. Gerhard, *The Plausibility of Life: Resolving Darwin's Dilemma* (New Haven and London: Yale University Press, 2005).

CHAPTER 5

1. Several sections of this chapter are excerpted from my article "Evolution, The Theory of" in the *Encyclopaedia Britannica*, Vol. 18 (2005): 855–891.

2. G. Mayr, B. Pohl, and D. S. Peters, "A Well-Preserved *Archaeopteryx* Specimen with Theropod Features," *Science* 310 (2005): 1483–1486.

3. E. B. Daeschler, N. H. Shubin, and F. A. Jenkins, Jr., "A Devonian Tetrapod-like Fish and the Evolution of the Tetrapod Body Plan," *Nature* 440 (2006): 757–763; N. H. Shubin, E. B. Daeschler, and F. A. Jenkins, Jr., "The Pectoral Fin of *Tiktaalik roseae* and the Origin of the Tetrapod Limb," *Nature* 440 (2006): 764–771.

4. Other features that are intermediate between reptiles and birds are the following: Unlike modern birds, *Archaeopteryx* had teeth and a long tail, similar in structure to that of the smaller dinosaurs. The hind legs are birdlike, but the well-preserved foot of the most recent specimen shows a hyperextensible second toe, similar to the killer claw of the dinosaur *Velociraptor*. The forelimbs retained primitive reptilian characteristics and had not yet completed their transformation into wings. *Archaeopteryx* may have been capable of flying, but it was not capable of sustained flight. The most recent specimen indicates that *Archaeopteryx* lived mostly on the ground, rather than in trees.

5. For example, these transitional fish still had gill cover plates, which pumped water over the gills, and the skull was fused to its shoulder girdle.

6. The morphological gap was also substantial because none of these animals (all between 75 centimeters and 1.5 meters in length) was truly an intermediate between fish and tetrapods.

7. The excavated fossils include three skulls, ten jawbones, and two specimens with head and trunk in one piece. *Tiktaalik* was a flattened, superficially crocodile-like animal, with skull about 20 centimeters in length. The pectoral fins are incipient forelimbs, with robust internal skeletons, but fringed with fin rays rather than digits. Fishlike features include small pelvic fins, the already-mentioned fin rays in their paired appendages, and well-developed gill arches, which suggest that they remained mostly aquatic. But the bony gill cover has disappeared, indicating reduced water flow through the gill chamber. The elongated snout suggests a shift from sucking up toward snapping up prey, mostly on land. The relatively large ribs indicate that *Tiktaalik* could support its body out of water.

CHAPTER 6

1. The fossil remains of *Homo floresiensis*, discovered in 2004 on the Indonesian island of Flores, seem related to *H. erectus*, although *H. floresiensis* is much smaller and lived around 12,000–18,000 years ago. These fossil remains are being actively investigated and their precise identification remains controversial.

2. T. D. White, G. WoldeGabriel, B. Asfaw et al., "Asa Issie, Aramis and the Origin of *Australopithecus*," *Nature* 440 (2006): 883–889.

3. DNA estimates of ancestral dates have broad ranges of possible variation (the so-called 95 percent "confidence interval") that may amount to as much as 30 percent of the estimate, several tens of thousands of years in the present case. As more and more DNA sequences are analyzed, the confidence interval decreases.

4. K. Owens and M.-C. King, "Genomic Views of Human History," *Science* 286 (1999): 451–453.

5. Some biologists, philosophers, and other people wonder whether dogs, whales, or monkeys might not have a sense of self, whether

they are self-aware just as we humans are. They assert that we don't know for certain, because we cannot know what is in the mind of non-human animals. This is of course correct, but I am persuaded that only humans have self-awareness on the basis of the following argument. Being self-aware implies being aware of one's own finitude, that our self will come to an end when we die. That is, self-awareness implies death-awareness. Death-awareness, in turn, calls for ceremonial burial of the dead. We treat other dead humans with respect, because we want to be so treated when we die. Humans are the only animals that ceremonially bury their dead. We are, I conclude, the only animals that are self-aware. This conclusion is, of course, congenial to people of faith who believe in the existence of the soul.

6. The mathematician and philosopher René Descartes (1596–1650) proposed that the soul influences the body by acting through the pineal gland of the brain. This fanciful suggestion leaves unresolved the issues that I have raised: which are the physiological correlates of the mental experiences. We know, for example, that memories are stored in the brain. However, our understanding of the brain states that encompass our mental experiences, or vice versa (i.e., understanding how our mental experiences impact our brain states), remains in its infancy.

CHAPTER 7

1. The ancestral organism, represented by the point from which the major branches first diverge, is sometimes referred to as LUCA, the Last Universal Common Ancestor. (See figure on page 81.)

2. A minor variation of the genetic code occurs in mitochondrial DNA.

3. Notice that the DNA of a gene is first "translated" into a related molecule, RNA (the so-called "messenger RNA") and that the genetic code refers to codons in the RNA.

4. The genetic information of some viruses is encoded in RNA, a molecule that consists also of four nucleotides, one of which

(represented as U, for uracil) is slightly different from the corresponding nucleotide in DNA (T). RNA is not generally organized as a double helix.

5. W. M. Fitch and E. Margoliash, "Construction of Phylogenetic Trees," *Science* 155(1967): 279–284.

Chapter 8

1. Aquinas, T. *Summa Theologiae* (see note 1 of Chapter 1).
2. M. Behe, *Darwin's Black Box* (see note 10 of Chapter 2).
3. W. Dembski, *The Design Inference: Eliminating Chance through Small Probabilities* (Cambridge, UK: Cambridge University Press, 1995).
4. P. Johnson, *The Wedge of Truth* (Downers Grove, IL: InterVarsity Press, 2000).
5. *Kitzmiller v. Dover Area School District*, 400 F. Supp. 2d 707 (M.D. Pa. 2005), slip opinion, 18; subsequent references are also to the slip opinion, which is available at http://www.pamd.uscourts.gov/kitzmiller/kitzmiller_342.pdf.
6. Ibid., 25, 28–29.
7. Ibid., 137.
8. Powerful microscopes have made it possible, in recent years, to examine materials at huge magnification, where the external configuration of presumptive atoms and molecules is observed. This is still a far cry from observing the atoms of current atomic theory, with their protons, neutrons, electrons, and other particles. It may very well be the case that scientists may eventually directly observe atoms and their detailed composition and configuration, but the atomic theory of the composition of matter does not depend on such observation. Nor would direct observation of atoms contribute much to the atomic theory, which indeed encompasses much more knowledge than simply claiming that atoms exist.
9. In the seventeenth century and beyond, most chemists accepted that every combustible substance, such as wood or coal, was in part composed of phlogiston. Burning was thought to be caused by the liberation of phlogiston, with the remaining substance left

as ash. Between 1770 and 1790, Antoine Lavoisier, the founder of modern chemistry, studied the gain or loss of weight when various substances were burned, and he demonstrated that phlogiston does not exist; eventually, he demonstrated that oxygen, which had been recently discovered by Joseph Priestly, was always involved in combustion and he went on to propose a general theory of oxidation.

10. *Kitzmiller v. Dover Area School District*, slip opinion, 5 (see note 5, above).

11. Behe, *Darwin's Black Box*, 39 (see note 10 of Chapter 2).

12. Ibid., 39.

13. Ibid., 72.

14. M. J. Pallen and N. J. Matzke, "From *The Origin of Species* to the origin of the bacterial flagella," *Nature Review Microbiology* 4 (2006): 784–790.

15. Ibid.

16. I. Musgrave, "Evolution of the Bacterial Flagellum," in *Why Intelligent Design Fails*, eds. M. Young and T. Edis (New Brunswick, NJ: Rutgers University Press, 2004), 48–84.

17. D. Ussery, "Darwin's Transparent Box. The Biochemical Evidence for Evolution," *Why Intelligent Design Fails*, eds. M. Young and T. Edis (New Brunswick, NJ: Rutgers University Press, 2004), 48–57.

18. K. Miller, "The Flagellum Unspun. The Collapse of 'Irreducible Complexity,'" in *Debating Design: From Darwin to DNA*, eds. W. Dembski and M. Ruse (Cambridge, UK: Cambridge University Press, 2004), 81–97.

19. Ibid., 85. A technical paper elaborating the evolutionary history of the genes that encode the components of the bacterial flagellum is: R. Liu and H. Ochman, "Stepwise Formation of the Bacterial Flagellar System," *Proceedings of the National Academy of Sciences USA* 104 (2007):7116–7121. The components of the flagellum are encoded by gene clusters that may include, in some species, upwards of 50 genes. The number of genes and the genes themselves greatly vary among different groups of bacteria. Liu and Ochman have identified all the flagellar proteins in 41 species from 11 quite diverse groups of bacteria. Twenty-four of the genes encoding the flagellar proteins were already present in the remote common ancestor of all

the bacterial species studied. The other genes have come about by duplication and evolution of preexisting genes. Moreover, many of the core of 24 ancestral genes are also derived from a few preexisting ones by successive gene duplications that gradually increased their number. The sequence similarity among all the flagellar genes in the 41 bacterial species has allowed Liu and Ochman to reconstruct the successive steps of addition and modification by which modern bacterial flagella have arisen.

20. Behe, *Darwin's Black Box*, 78 (see note 10 of Chapter 2).

21. Ibid., 97.

22. *Thrombosis Haemostasis* 70 (1993): 24–28.

23. Miller, *Finding Darwin's God*, 152-161. Additional non-technical discussions can be found in B. H. Weber and D. J. Depew, "Darwinism, Design, and Complex Systems Dynamics," in *Debating Design: From Darwin to DNA*, eds. W. Dembski and M. Ruse (Cambridge, UK: Cambridge University Press, 2004), 173–190. Also see N. Shanks and I. Karsai, "Self-Organization and the Origin of Complexity," in *Why Intelligent Design Fails*, eds. M. Young and T. Edis (New Brunswick, NJ: Rutgers University Press, 2004), 85–106; Miller, "The Flagellum Unspun" (see note 18, above).

24. Behe, *Darwin's Black Box*, 185 and 138 (see note 10 of Chapter 2). In fact, examples of complex biochemical structures or systems that have arisen from simpler components are very, very numerous. One that has appeared in a scientific journal I received August 28, 2006, while working on this manuscript, concerns the phylogenetic enigma of snail hemoglobin. It turns out that the complex hemoglobin of the planorbid snail *Biomphalaria glabrata* has evolved from pulmonate myoglobin by a simple evolutionary mechanism that creates a high molecular mass respiratory protein from seventy-eight similar globin domains. See B. Lieb, K. Dimitrova, H.-S. Kang et al., "Red Blood with Blue-Blood Ancestry: Intriguing Structure of a Snail Hemoglobin," *Proceedings of the National Academy of Sciences USA* 103 (2006): 12011–12016. An example that involves the evolution of a particular protein from a single-cell ancestor to animal descendants appears in the same journal issue: Y. Segawa, H. Suga, N. Iwabe, C. Oneyama, T. Akagi, T. Miyata, and M. Okada, "Functional Development of Src Tyrosine

Kinases During Evolution from a Unicellular Ancestor to Multi-cellular Animals," *Proceedings of the National Academy of Sciences USA* 103 (2006): 12021–12026.

25. *Kitzmiller v. Dover Area School District*, slip opinion, 78–79 (see note 5, above). Behe is also inclined to exaggerate claims about the significance of his own accomplishments: his discovery of intelligent design in biochemistry "is so unambiguous and so significant that it must be ranked as one of the greatest achievements in the history of science," *Darwin's Black Box*, 232.

26. Th. Dobzhansky, *The Biology of Ultimate Concern* (New York: New American Library, 1967), 13.

27. Dembski, *The Design Inference*, 3 (see note 3, above).

28. Behe, *Darwin's Black Box*, 223 (see note 10 of Chapter 2).

29. R. T. Pennock, *Tower of Babel. The Evidence Against the New Creationism* (Cambridge, MA: MIT Press, 2002).

30. S. E. Lawrence, "Sexual Cannibalism in the Praying Mantis, *Mantis religiosa*: A Field Study," *Animal Behaviour* 43 (1992): 569–583; see also M. A. Elgar, "Sexual Cannibalism in Spiders and Other Invertebrates," in *Cannibalism: Ecology and Evolution among Diverse Taxa*, eds. M. A. Elgar and B. J. Crespie (Oxford: Oxford University Press, 1992).

31. J. A. Downes, "Feeding and Mating in the Insectivorous Ceratopogoninae (Diptera)," *Memoirs of the Entomological Society of Canada* 104 (1978): 1–62.

32. Diverse sorts of oddities associated with mating behavior, such as those I have just described, are detailed in the delightful, but accurate and well-documented, book by Olivia Judson, *Dr. Tatiana's Sex Advice to All Creation* (New York: Holt, 2002). Perplexing and even bizarre features in the configuration and behavior of all sorts of organisms appear in John C. Avise's *Evolutionary Pathways in Nature. A Phylogenetic Approach* (Cambridge, UK: Cambridge University Press, 2006), a scholarly book about the molecular evolutionary history of diverse organisms and their characteristics. Take *male* pregnancy in fish as a startling example. In the 200-plus species of the family Syngnathidae, which includes pipefishes and seahorses, the males rather than the females bear the developing embryos in a pouch located under the male's abdomen. The female

transfers her unfertilized eggs to the male's pouch, where they are fertilized by his sperm. The male then carries the developing embryos for several weeks, eventually giving birth to the offspring, which are miniature versions of the adults. During his pregnancy, the father nourishes and protects the brood, while the mother plays no role. In some species, the female produces far more eggs than can be accommodated within a male's pouch and, consequently, females are polygamous. See A. B. Wilson, I. Ahnesjö, A. Vincent, and A. Meyer, "The Dynamics of Male Brooding, Mating Patterns, and Sex Roles in Pipefishes and Seahorses (Family Syngnathidae)," *Evolution* 57 (2003): 1374–1386.

33. F. Darwin (ed.), *The Life and Letters of Charles Darwin,* 3 vols (London: Murray, 1887), vol. 2, 105.

34. D. L. Hull, "God of the Galapagos," *Nature* 352 (1992): 485–486.

35. D. Hume, *Dialogues Concerning Natural Religion,* ed. N. K. Smith (Oxford: Oxford University Press, 1935), 244. This formulation of the problem of evil in God's world seems to be due to the philosopher of Classical Greece, Epicurus (341–270 B.C.). Voltaire, in his dictionary article on *Bien* ("Goodness") quotes Epicurus, according to Lactantius: "Either God can remove evil from the world and will not; or being willing to do so, cannot. . . . If he is willing and cannot, he is not omnipotent. If he can but will not, he is not benevolent. . . . If he both wants and can, whence comes evil over the face of the earth?" See François-Marie Arouet de Voltaire, *The Encyclopedia of Philosophy,* vol. 8 (London: Macmillan, 1967), 262–270. The eminent evolutionist Dobzhansky wrote: "If the universe was designed to advance toward some state of absolute beauty and goodness, the design was incredibly faulty. . . . Why so many false starts, extinctions, disasters, misery, anguish, and finally the greatest of evils—death? The God of love and mercy could not have planned all this. Any doctrine which regards evolution as predetermined or guided collides head-on with the ineluctable fact of the existence of evil." *The Biology of Ultimate Concern,* 120 (see note 26, above).

36. J. Haught, "Darwin's Gift to Theology," in *Evolutionary and Molecular Biology: Scientific Perspectives on Divine Action,* eds. R. J. Russell, W. R. Stoeger, and F. J. Ayala (Vatican City State and Berkeley, CA:

Vatican Observatory and the Center for Theology and the Natural Sciences, 1998), 393–418.

37. A. Peacocke, "Biological Evolution—A Positive Appraisal," in *Evolutionary and Molecular Biology: Scientific Perspectives on Divine Action,* eds. R. J. Russell, W. R. Stoeger, and F. J. Ayala (Vatican City State and Berkeley, CA: Vatican Observatory and the Center for Theology and the Natural Sciences, 1998), 357–376.

38. See several other contributions to *Evolutionary and Molecular Biology: Scientific Perspectives on Divine Action.*

CHAPTER 9

1. C. Hodge, *What Is Darwinism* (New York: Scribner, Armstrong & Co., 1874), 48–51.

2. A. H. Strong, *Systematic Theology,* 3 vols (Westwood, NJ: Fleming Revell, 1907), vol. 2, 472–473.

3. The address of Pope John Paul II to the members of the Pontifical Academy of Sciences appeared in *L'Osservatore Romano* on October 23, 1996, in its French original, and on October 30, 1996, in English. Both texts are reproduced in *Evolutionary and Molecular Biology: Scientific Perspectives on Divine Action,* eds. R. J. Russell, W. R. Stoeger, and F. J. Ayala (Vatican City State and Berkeley, CA: Vatican Observatory and the Center for Theology and the Natural Sciences, 1998), 2–9.

4. *Voices for Evolution* (rev. ed.), ed. Molleen Matsumura (Berkeley, CA: National Center for Science Education, 1995, 176 pp) contains a large collection of statements in support of evolution made by scientific, religious, educational, and civil liberties organizations. The two quotations are on pages 107 and 96, respectively. A good source for documents, as well as an excellent discussion of the controversy, is E. C. Scott, *Evolution vs. Creationism* (Westport, CT: Greenwood Press, 2004), *xxiv* + 272 pp.

5. Augustine, *The Literal Meaning of Genesis* (*De Genesi ad Litteram*), Book 2, Chap. 9. There are several translations of this book into English: see, for example, the edition "Translated and Annotated

by John Hammond Taylor, S.J.," 2 volumes (New York: Newman Press). The quotation is from vol. 1, 58–59.

6. Augustine, *De Genesi ad Litteram*, Book 1, Chapter 17, vol. 1, 38–40 (see note 5, above).

7. *Voices*, 97 (see note 4, above).

8. *McLean v. Arkansas Board of Education* 529 F. Supp. 1255, 1982. I testified as an expert witness at the trial in Little Rock.

9. *Edwards v. Aguilar* 482 U.S.578, 1987. An *amicus brief* that the U.S. National Academy of Sciences introduced in the Supreme Court proceedings incorporates the booklet "Science and Creationism. A View from the National Academy of Sciences," of which I was a principal author, although unnamed, except as a member of the committee that drafted the document.

10. D. J. Futuyma, *Evolutionary Biology*, 3rd ed. (Sunderland, MA: Sinauer, 1998), 5. It seems that, upon further reflection, Futuyma has changed his views: "But does evolutionary biology deny the existence of a supernatural being or a human soul? No, because science, including evolutionary biology, is silent on such questions. By its very nature, science can entertain and investigate only hypotheses about material causes that operate with at least probabilistic regularity. It cannot test hypotheses about supernatural beings or their intervention in natural events." *Evolution* (Sunderland, MA: Sinauer, 2005), 12.

11. R. Dawkins, *River Out of Eden* (New York: HarperCollins, 1992), 133.

12. W. Provine, "Evolution and the Foundation of Ethics," *MBL Science* 3 (1988): 25–29.

13. National Academy of Sciences, *Teaching About Evolution and the Nature of Science* (Washington, DC: National Academy Press, 1998), 58.

14. Ibid.

15. K. Miller, *Finding Darwin's God* (New York: HarperCollins, 1999); J. Haught, "Darwin, Design, and Divine Providence," in *Debating Design: From Darwin to DNA*, eds. W. Dembski and M. Ruse (Cambridge, UK: Cambridge University Press, 2004), 229–245; Haught, "Darwin's Gift to Theology" (see note 36 of Chapter 8); Peacocke, "Biological Evolution" (see note 37 of Chapter 8). See also I. G.

Barbour, "Five Models of God and Evolution," in *Evolutionary and Molecular Biology: Scientific Perspectives on Divine Action,* eds. R. J. Russell, W. R. Stoeger, and F. J. Ayala (Vatican City State and Berkeley, CA: Vatican Observatory and the Center for Theology and the Natural Sciences, 1998), 419–442; and other contributions to that volume.

16. *De Genesi ad Litteram,* 1, 19 (see note 5, above). Further in this chapter, Augustine writes:

> Now, it is a disgraceful and dangerous thing for an infidel to hear a Christian, presumably giving the meaning of Holy Scripture, talking nonsense on these topics [the Earth, the heavens, the motion and orbit of the stars, the kinds of animals and shrubs]; and we should take all means to prevent such an embarrassing situation, in which people show up vast ignorance in a Christian and laugh it to scorn. The shame is not so much that an ignorant individual is derided, but that people outside the household of the faith think our sacred writer held such opinions, and, to the great loss of those for whose salvation we toil, the writers of our Scripture are criticized and rejected as unlearned men. If they find a Christian mistaken in a field which they themselves know well and hear him maintaining his foolish opinions about our books, how are they going to believe those books in matters concerning the resurrection of the dead, the hope of eternal life, and the kingdom of heaven, when they think their pages are full of falsehoods on facts which they themselves have learnt from experience and the light of reason? Reckless and incompetent expounders of Holy Scripture bring untold trouble and sorrow on their wiser brethren when they are caught in one of their mischievous false opinions and are taken to task by those who are not bound by the authority of our sacred books. For then, to defend their utterly foolish and obviously untrue statements, they will try to call upon Holy Scripture for proof and even recite from memory many passages which they think support their position, although *they understand neither what they say nor the things about which they make assertion.*

Ibid., 42–43 (italics in the original).

17. Camus also wrote: "I know that something in this world has a meaning and this is man; because he is the only being that demands to have a meaning." The literary critic Erich Auerbach (1892–1957) wrote in his 1929 *Mimesis: The Representation of Reality in Western Literature* that "serious treatment of everyday reality" requires "the broad and elastic form of the novel."

18. L. Eiseley, *The Man Who Saw Through Time* (New York: Charles Scribner's Sons, 1972), 103.

19. F. Dyson, "Religion from the Outside," *The New York Review* (June 22, 2006), 4–8.

CHAPTER 10

1. J. S. Mill's best known philosophical work is *System of Logic* (1843). His theory of knowledge was further developed in his *Examination of Sir William Hamilton's Philosophy* (1865).

2. F. Darwin, *More Letters of Charles Darwin*, 2 volumes (London: Murray, 1903), vol. 1, 195.

3. See F. J. Ayala, "On the Scientific Method, Its Practice and Pitfalls" *History and Philosophy of the Life Sciences* 16 (1994): 205–240.

4. K. R. Popper, *The Logic of Scientific Discovery* (London: Hutchinson, 1959).

5. These two "episodes" have recently been described, particularly by molecular biologists, as "discovery-based science" and "hypothesis-driven science," respectively. Discovery-based science is at times associated with "big science" and the work of interdisciplinary research teams, whereas hypothesis-driven science is said to be "small science," primarily carried out by individual investigators. In different proportions or degrees, science always involves both discovery and testing.

6. C. Darwin, *The Autobiography of Charles Darwin (1809–1882)*, ed. N. Barlow (London: Collins, 1958).

7. See P. B. Medawar, *The Art of the Soluble* (London: Methuen, 1967). This small book provides a very eloquent, yet profound, discussion of the scientific method as a dialogue between the two essential

episodes of science: conjectures and refutations. My discussion of this subject is importantly derived from Medawar's.

8. This is how molecular biologist François Jacob, in *The Statue Within* (New York: Harcourt, Brace and World, 1964) describes the process:

> What had made possible analysis of bacteriophage multiplication, and understanding of its different stages, was above all the play of hypotheses and experiments, constructs of the imagination and inferences that could be drawn from them. Starting with a certain conception of the system, one designed an experiment to test one or another aspect of this conception. Depending on the results, one modified the conception to design another experiment. And so on and so forth. That is how research in biology worked. Contrary to what once I thought, scientific progress did not consist simply in observing, in accumulating experimental facts and drawing up a theory from them. It began with the invention of a possible world, or a fragment thereof, which was then compared by experimentation with the real world. And it was this constant dialogue between imagination and experiment that allowed one to form an increasingly fine-grained conception of what is called reality (pp. 224–225).

9. One example is the replacement of Newtonian mechanics by the theory of relativity, which rejects the conservation of matter and the simultaneity of events that occur at a distance—two fundamental tenets of Newton's theory. Examples of this kind are pervasive in rapidly advancing disciplines, such as molecular biology. The so-called "central dogma" held that molecular information flows only in one direction, from DNA to RNA to protein. The DNA contains the genetic information that determines what the organism is, but that information has to be expressed in enzymes (a particular class of proteins) that guide all chemical processes in cells. The information contained in the DNA molecules is conveyed to proteins by means of intermediate molecules, called messenger RNA. Molecular biologists David Baltimore and Howard Temin were awarded the Nobel Prize for discovering that information could flow in the opposite direction, from RNA to DNA, by means of the

enzyme reverse transcriptase. They showed that some viruses, as they infect cells, are able to copy their RNA into DNA, which then becomes integrated into the DNA of the infected cell, where it is used as if it were the cell's own DNA (D. Baltimore, "Viral RNA-Dependent DNA Polymerase in Virions of RNA Tumor Viruses," *Nature* 226 [1970]: 1209–1211; H. M. Temin and S. Mizutani, "RNA-Dependent DNA Polymerase in Virions of Rous Sarcoma Virus," *Nature* 226 [1970]: 1211–1213).

Another example is the following: For many years, it was universally thought that only the proteins known as enzymes could mediate (technically "catalyze") the chemical reactions in cells. However, molecular biologists Thomas Cech and Sidney Altman received the Nobel Prize in 1989 for showing that certain RNA molecules act as enzymes and catalyze their own reactions (T. R. Cech, "Self-splicing RNA: Implications for Evolution," *International Review of Cytology* 93 [1985]: 3–22).

These revolutionary hypotheses were published after their authors had subjected them to severe empirical tests. Theories that are inconsistent with well-accepted hypotheses in the relevant discipline are likely to be ignored when they are not availed by convincing empirical evidence. The microhistory of science is littered with farfetched or ad hoc hypotheses, often proposed by individuals with no previous or posterior scientific achievements. Theories of this sort usually fade away because they are ignored by most of the scientific community, although on occasion they engage the community's interest because the theory may have received attention from the media or even from political or religious bodies. The flop a few years ago over "cold fusion" is an example of an unlikely and poorly tested hypothesis that received some attention from the scientific community because its proponents were well-established scientists. (The hapless protagonists of the cold fusion fiasco are Martin Fleishmann and B. Stanley Pons. The tale is well told by G. Taubes, *Bad Science: The Short Life and Weird Times of Cold Fusion* [New York: Random House,1993]).

10. Popper, 1959, *The Logic of Scientific Discovery;* see also K. R. Popper, *Conjectures and Refutations: The Growth of Scientific Knowledge*

(London: Routledge and Kegan Paul, 1963); G. C. Hempel, *Aspects of Scientific Explanation* (New York: Free Press, 1965).

11. C. Darwin, *Autobiography,* 119.

12. C. Darwin, *On the Origin of Species* (see note 1 of Chapter 3), 1 (my italics).

13. C. Darwin, *Autobiography,* 109 (my italics).

14. F. Darwin, *More Letters of Charles Darwin,* vol. 2, 323 (italics in the original). See D. Hull, *Darwin and His Critics* (Cambridge, MA: Harvard University Press, 1973).

15. F. Darwin, *More Letters of Charles Darwin,* vol. 1, 195.

16. C. Darwin, *Autobiography,* 141.

17. See, e.g., G. De Beer, *Charles Darwin, A Scientific Biography* (Garden City, NY: Doubleday, 1964); M. T. Ghiselin, *The Triumph of the Darwinian Method* (Berkeley, CA: University of California Press, 1969); Hull, *Darwin and His Critics*; E. Mayr, "Introduction" in C. Darwin, *On The Origin of Species* (see note 1 of Chapter 3). See also Ayala, "On the Scientific Method" (note 3, above).

18. C. Darwin, "Darwin's Notebooks on Transmutation of Species," ed. G. De Beer, *Bulletin of the British Museum (Natural History)* 2 (1960): 23–200 (especially page 142).

19. C. Darwin, *The Descent of Man and Selection in Relation to Sex* (London: Murray, 1871; 2nd ed., 1889), 606. See also note 9, above.

20. Mendel's paper, written in German, has been reprinted in English translation in numerous publications. The one I quote is from E. W. Sinnot, L. C. Dunn, and Th. Dobzhansky, *Principles of Genetics* (New York: McGraw-Hill, 1958), Appendix, 419–443. A short biography of Mendel, as well as an annotated edition of his classic paper, can be found in A. F. Corcos and F. V. Monaghan, *Gregor Mendel's Experiments on Plant Hybrids. A Guided Study* (New Brunswick, NJ: Rutgers University Press, 1993). An authoritative biography is V. Orel, *Gregor Mendel, the First Geneticist* (Oxford: Oxford University Press, 1996).

21. Darwin's tentative explanation of biological heredity is presented in the second volume of his *The Variation of Animals and Plants under Domestication* (see note 3 of Chapter 4, above).

22. Vítezslav Orel, *Gregor Mendel, The First Geneticist* (Oxford: Oxford University Press, 1996), 191.

23. Ibid., 191–199.
24. This difficulty was famously pointed out in 1867 by Fleeming Jenkin, a professor of engineering, in *The North British Review*. Darwin had anticipated in 1838, in his Transmutation Notebook C, the problem emerging from blending inheritance. But no alternative was known to Darwin even in 1868. He tamely, and unpersuasively, sought to escape the problem by distinguishing between variations that appear in a single individual and those that might appear simultaneously in many individuals of the population. The latter would allow for matings between individuals carrying the advantageous variant, thus avoiding the dilution due to blending inheritance.
25. R. L. Numbers, *Darwinism Comes to America* (Cambridge, MA, and London: Harvard University Press, 1998); P. J. Bowler, *The Non-Darwinian Revolution. Reinterpreting a Historical Myth* (Baltimore and London: Johns Hopkins University Press, 1988).
26. The quotation comes from the last paragraph of Darwin's *The Origin*, 1st ed., 490.

ACKNOWLEDGMENTS

This book is a much expanded version of my *Darwin and Intelligent Design* (Minneapolis: Fortress Press, 2006), which was written at the request of Michael West, editor-in-chief of Fortress Press. Mr. West had asked me to consider crafting a short book for intelligent laypersons in which I would explain the status and role of evolutionary theory with respect to the assertions made by proponents of intelligent design. I had previously promised the editors of the Joseph Henry Press that I would write a book on the same subject, although directed to a different audience. I am grateful that Mr. West and the principals of JHP agreed that the two projects were compatible.

For their assistance, sensitivity, and wisdom I am much indebted to Barbara Kline-Pope, director; Stephen Mautner, executive editor; and Dick Morris, managing editor, of the Joseph Henry Press and the National Academies Press. My gratitude extends to Jeffrey Robbins, who copyedited the manuscript. My debt to Denise Chilcote for preparing the manuscript is very extensive and so is my gratitude. As my executive assistant for two decades, she has never failed to do whatever needed to be done as well as it could be done. I have tremendously benefited from her dedication and perfectionism.

INDEX